Annual Conference of the American Forage and Grassland Council (AFGC 2023)

Winston-Salem, North Carolina, USA
8-11 January 2023

ISBN: 978-1-7138-8313-5

Printed from e-media with permission by:

Curran Associates, Inc.
57 Morehouse Lane
Red Hook, NY 12571

Some format issues inherent in the e-media version may also appear in this print version.

Printed by Curran Associates, Inc. (2024)

For permission requests, please contact American Forage and Grassland Council (AFGC) at the address below.

American Forage and Grassland Council (AFGC)
PO Box 867
Berea, KY 40403
USA

Phone: 800.944.2342

www.afgc.org

Additional copies of this publication are available from:

Curran Associates, Inc.
57 Morehouse Lane
Red Hook, NY 12571 USA
Phone: 845-758-0400
Fax: 845-758-2633
Email: curran@proceedings.com
Web: www.proceedings.com

TABLE OF CONTENTS

ALTERNATIVE METHODS FOR DETERMINING FORAGE MINERAL CONTENT USING NEAR-INFRARED SPECTROSCOPY INSTRUMENTATION

D. McIntosh[1], C. Sams[1], D. Butler[1], G. Bates[1], R. Florence[2], B. Husmoen[3], and R. Lemus[4]

Forages contain macro and micro minerals that are often not accounted for due to expensive wet chemistry testing, or the distrust in near infrared spectroscopy (NIRS) predictions. Until advances in NIRS instrumentation were made over the last decade, it was not possible for calibrations to be developed with accuracy and validation necessary to change this. Variability in forages has often kept mineral calibration development secondary to the basic nutritive analysis components. This project brings new techniques at the sample presentation level and alternatives to sampling for wet chemistry analysis that is more representative of the spectra used when creating a calibration set. Traditional sampling methods used in quantitative analysis present problems where the actual sample used for wet chemistry is often a sub-sample not related to the spectra collected from the same sample. Thus, when calibrations were developed, wet chemistry analysis data does not match the spectra collected. This project provides new possibilities to develop and expand NIRS calibrations with accuracy not seen before in forages. Predictions for macro minerals Phosphorus (P), Potassium (K), Sulfur (S), Calcium (Ca), Magnesium (Mg) and, micro-minerals of Sodium (Na), Manganese (Mn), Selenium (Se), Copper (Cu), Boron (B), Zinc (Z), and Iron (Fe) are now possible with varied success. This feasibility study examined three harvest years of Tall Fescue and Orchardgrass from Tennessee producers with results using the techniques presented and calibration validation.

[1]Department of Plant Sciences, The University of Tennessee, Knoxville; [2] Plant, Pest and Soil Center, Nashville, TN; [3]Near Infrared Spectroscopy Forage and Feed Consortium, Hillsboro, WI; [4]Mississippi State University, Mississippi State, MS.

BLOOMING AND FORAGE CHARACTERISTICS OF TWELVE NATIVE FORBS AND LEGUMES SUBJECTED TO REPEATED DEFOLIATION

Jessica Prigge[1,2], Eric Bisangwa[1], and Patrick Keyser[1]

Pastures provide an opportunity to serve both growing beef cattle and critical pollinators by including a diverse set of native forbs and legumes. However, native forbs and legumes have not been evaluated for various forage characteristics. Therefore, the objectives of this study were to examine forage biomass, blooming patterns, and nutritive composition of 12 native forb and legume species in a complete, randomized block design, small plot study. Twelve species were seeded in 2018 and monitored between 2020 and 2022. Analysis was conducted using R software (version 4.1.1) and statistical significance was set at $P \leq 0.05$. Treatments were compared for forage biomass, blooming periods, CP, NDF, and ADF using a mixed model ANOVA. Forage biomass varied among species ($P<0.001$) but was greatest for cup plant (3,500-7,100 lb/ac) and Maximilian sunflower (700-4,300 lb/ac) per harvest. Each species exhibited varying blooming periods. Black-eyed Susan and lanceleaf coreopsis bloomed earlier in the summer, purple coneflower and oxeye sunflower bloomed during mid-summer, and both cup plant and Maximillian sunflower bloomed in late summer and fall. When including a diversity of forbs, the variety of blooming periods could sustain pollinators throughout the growing season. Most species met or exceeded the nutritional needs of a growing beef steer (>15% crude protein), suggesting that they could be beneficial in a pasture system.

[1]Center for Native Grasslands Management, School of Forestry, Wildlife, and Fisheries, University of Tennessee, Knoxville, TN; [2]Presenting Author, jprigge@vols.utk.edu.

BUILDING PARTNERSHIPS TO ADDRESS SOCIAL AND TECHNOLOGICAL CHALLENGES TO ENHANCE FARM PROFITABILITY AND IMPROVE WATER QUALITY THROUGH BETTER GRASSLAND MANAGEMENT

Stafford, C., Culpeper County, Clark, R.A., Shenandoah County, Ritchie, L., VT Sociology, Dudding, J. Craig County, Pent, G., VT SVAREC, Fike, J., VT VCE, Benner, J., Augusta County, Swanson, C., Albemarle County, Baker, S., Bedford County, Mize, T., Fauquier County, Temu, V., VSU VCE, Payne, K., VT SPAREC, Gill, D., Mullins, D. VT Sociology, McGuire, R., VT VCE, Teutsch, C., UK, Thomason, W., VT VCE, Grev, A., UM, Blevins, P., VCE Washington, Clarke, T., VCE Mecklenburg, Poore, M., NC State, Booher, M., VCE Rockingham, Stanley, T.,VCE Rockbridge, Halich, G., UK, Bovay, J.,VT VCE, Jones, S., John Marshall Soil and Water Conservation District, Love, K., VCE Rappahannock, Hilliard, A., VSU, Byington, A., VCE Lee, Mullins, E., VCE Page, Haugen, I., VT Library.

Abstract

With 2.1 million acres of pastureland and 1.25 million acres of hay land in Virginia, the rural Virginia landscape is predominately grassland. These lands form the base of the $3.96 billion-dollar livestock and dairy industry in Virginia. Managing these livestock in a profitable manner for farmers and beneficial to the environment is important. A cultural tradition with roots in colonial times has been to run animals in large fields year-round throughout Virginia. Livestock often graze from spring until fall (about 220 days), and farmers feed hay the remainder of the year. Spikes in the cost of fuel, fertilizer, and equipment are making traditional grazing/haying systems less profitable. The Virginia Cooperative Extension Farm Enterprise budgets show that that the cost of hay accounts for over 50% of the cost of sustaining livestock annually. University of Kentucky shows that most cow-calf producers maximize their profitability by shifting from grazing 220 days to grazing 275 to 300 days. Extension agents working with livestock producers found that they could improve their profitability by at least $75 per cow by extending their grazing season. The same phenomenon applies to other types of grazing livestock. If ten percent of the livestock producers in the state adopted better grazing management to extend their grazing season by 60 days, profitability is expected to for Virginia grazing livestock producers by over $5 million per year. Practices such as rotational grazing and stream exclusion are directly tied to National and State goals to improve water quality in the Chesapeake Bay. Virginia's Phase III WIP (Chesapeake Bay Watershed Improvement Plan) seeks the exclusion of livestock from all perennial streams and achieving good rotational grazing practices on 347,000 acres of pasture. A number of agencies and private sector groups have been providing cost share and technical guidance to incentivize livestock stream exclusion and the installation of pasture management infrastructure. Installation is only part of the challenge. Farmers also need to be taught how to how to manage the system in a profitable manner and have been slow to adopt good pasture management practices. Preliminary data show that 87% of Virginia's cow-calf producers manage their grasslands using traditional methods. Only six percent have extended their grazing season beyond 265 days.

Introduction, Action Plan, Methods and Results:

Objective #1/Action Taken: Determine the Social factors that influence farmer change, particularly in the context of forage-based livestock systems.

Survey Developed: Social Science Research on Factors Influencing Farmers' Attitudes Toward the Adoption of Rotational Grazing in Virginia

From this we evaluated various factors that influence farmers' attitudes toward the adoption of rotational grazing. Qualitative and quantitative data gathered from a convenience sample of Virginia farmers. 535 completed surveys received from; farmers practicing continuous grazing (206) from farmers practicing rotational grazing (329). Three barriers found for both types of producers identified as a significant or moderate challenge: amount of work required to start rotational grazing, shade, and water source constraints. Continuous grazers said they get more done with machinery, rather drive than walk, and like using hay machinery as top reasons for not adopting rotational grazing.

Objective #2/Action Taken: Improve the competency of all partner agencies and agribusiness by providing in-depth grazing management training.

A four-month long training series delivered via Canvas was designed to teach professionals (Extension Agents, Agency Representatives, and Agribusiness Employees) about grazing management. There were 118 participants in 2021, and now a second cohort of 52 participants are enrolled in this course in 2022. There were 23 different presentations that provided over 8 hours of technical training offered in the online course. This training program is being offered again at the request of Virginia State University and other organizations.

Thirty-six professionals participated in hands-on workshops held at three Virginia AREC's. The six-hour training provided participants with hands-on, in the field training in grazing management contributing to CALS strategic plan with other universities and stakeholders.

Objective #3/Action Taken: Expand the Graze 300 VA technical resource base.
Develop and assimilate technical resources filling gaps with new videos, new publications and tools professionals and producers can use in addressing needs.

Publications Written:
Managing the Spring Pasture Flush, author Matt Booher, under review, pending publication.
Adding Legumes to Pasture, author John Fike, used in training agricultural retailers.

Videos Produced:
Six educational videos were produced highlighting key aspects of extended grazing practice for livestock. Interviews conducted with cattle and small ruminant producers were used to discuss issues like: drought, spring flush, intensive rotational grazing, electric fencing, and more. Two

videos focused on fence chargers and temporary fence options with discussions on accommodations made due to age/type of animal, or predator, terrain and weather issues. These new videos will be posted to the Graze 300 website referenced below.

Poster Developed:
"Building Partnerships to Address Social and Technological Challenges to Enhance Farm Profitability and Improve Water Quality Through Better Grassland Management." Presented at the annual meeting of the Center for Advanced Innovation in Agriculture, Virginia Tech University, March 28, 2022.

Website Improved:
Graze 300 VA Web site https://ext.vt.edu/agriculture/graze-300.html experienced 1,986 pageviews over the past year, 8 existing videos were viewed 22,000 times and new videos are slated for this site.

Instruments Built:
75 falling plate meters built following the Ed Rayburn, WVU design. These were distributed to professionals for their use in training farmers how to measure stockpiled forage inventory.

Objective #4: Broaden the use of technology to further advance the Graze 300 VA Initiative.

Theory Developed:
Recordkeeping is identified as important to grazing livestock production and as VCE budgets put harvested feed at some 60% of total expense, an economic opportunity is found. The use of digital is thought to improve convenience and ease of use by producers.

A review of current smartphone and tablet applications and computer programs was conducted. PastureMap® was selected as the program that would be the best fit to pilot with Virginia farmers beginning April 2022. 11 pilot farms were identified. Use of the software confirmed producer's management action taken, meaning it was accurate by comparison.

Acknowledgements:
The Graze 300 VA team wishes to thank both sources of funding for this project including: VT CALS Strategic Plan Advancement, 2021 Integrated Internal Competitive Grants Program Center for Advanced Innovation in Agriculture, SmartFarm Innovation Network TM and Southern SARE Professional Development Grant.

References:
Mullins, D., et al, 2022. Social science research on factors influencing farmers attitudes toward the adoption of rotational grazing in Virginia. https://ext.vt.edu/content/dam/shenandoah_ext_vt_edu/files/ag/graze300/09-20-22%20FINAL%20GRAZE%20300%20Social%20Dimensions%20Report.pdf

Graze 300 VA, 2022. Stafford, C., et al. https://ext.vt.edu/agriculture/graze-300.html

USING RED-GREEN-BLUE IMAGERY TO QUANTIFY COOL AND WARM SEASON GRASSES IN PASTURES

J.A. Bush[1], C.D. Teutsch[2], S.R. Smith[1], and J.C. Henning[1]

Abstract

Anecdotal evidence suggests that warm season grasses are increasing in tall fescue based pastures in the transition zone states such as Kentucky. This may impact how pastures need to be managed to optimize production. The objective of this study was to evaluate Red-Green-Blue imagery for quantifying cool (C3) and warm season (C4) grass ratios in pastures. In the fall of 2020 and spring of 2021, imagery was captured using a digital-SLR camera mounted on a 10 ft extension pole and a drone equipped with a 20 megapixel camera on seven farms in western Kentucky and one farm in western Tennessee. Two fields were sampled on each farm. Images were then analyzed using a free green canopy cover (GCC) analyzing software called Canopeo (OSU, Stillwater). The GCC estimates were compared to reference values obtained using occupancy grids. Drone images provided the best estimate of green canopy cover in the fall, but not the spring. Residual values were 12 and 22% and 14 and 5% for the drone and camera in the spring and fall, respectively. Green canopy cover software could estimate the proportion of C3 grasses and C4 in images captured by a drone and a Nikon camera with reasonable accuracy. More work is needed to confirm these results.

Introduction: Kentucky farmers generally rely on cool-season forages such as tall fescue (*Schedonorus arundinaceus* (Schreb.) Dumort., nom. cons.), orchardgrass (*Dactylis glomerata* L.), timothy (*Phleum pratense* L.), and Kentucky bluegrass (*Poa pratensis* L.) (Lacefield, 2014). Some warm-season grasses are used in Kentucky forage systems however they cannot be the sole source of forage due to their short period of productivity that is limited to the warmest parts of the year, June-August (Ball et al., 2007). Anecdotal evidence suggests that warm season grasses are increasing in tall fescue-based pastures in the transition zone states such as Kentucky. This may impact how pastures need to be managed to optimize production. The objective of this study was to evaluate Red-Green-Blue imagery for quantifying cool (C3) and warm season (C4) grass ratios in pastures.

Materials and Methods: This study was conducted on seven farms across western Kentucky and one farm in northwestern Tennessee. Farms were in Caldwell (3), Trigg (1), Fulton (1), Marshall (1), and Daviess (1) counties in Kentucky and Weakley County (1) in Tennessee. Two fields per farm were sampled. Median field size was 10 ac with a range of 3-15 acres.

Digital images were captured using a Nikon® D750 digital single-lens reflex (DSLR) camera with an AF-S NIKKOR 20mm f/1.8G ED Wide-Angle Lens and a ProMaster ® 77 mm UV Haze Ultraviolet Filter. This camera was equipped with a 24.3MP CMOS (complementary metal oxide semiconductor) sensor. A DocaDone DocaPole, 6 – 24 Foot Extension Pole with a ClickSnap Camera Swivel Adapter, was used to suspend the camera approximately 10 ft above the ground. Photos were captured remotely using the Nikon Wireless Mobile Utility application for iOS. Camera settings were as follows: bit depth = 24, image quality of NEF + Jpeg (Fine), jpeg compression = optimum, automatic white balance, f/11 aperture, 200 ISO. Thirty images per field were captured and samplings were done in the fall of 2020 and spring of 2021. The timing of all pasture sampling was after one of more frost events (<0 °C) that resulted in senescence of warm season grasses. The timing in the spring was before significant green-up had occurred for warm season grasses. Images were captured every twenty steps while walking in a random zig-zag pattern. The start and end times were recorded and used to calculate the time required.

Aerial images were captured using a DJI Phantom 4 Pro V2.0 quadcopter with a 1-inch CMOS sensor (20MP). FieldAgent by Sentera (Saint Paul, Minnesota) was used for flight planning and execution. Images were uploaded to DroneDeploy (San Francisco, CA) for orthomosaic stitching. Drone flights were done at 50-100' altitude, 6-8 mph, and 40% overlap to maximize resolution and optimize battery life. Drone flights were executed over each field in the fall of 2020 and the spring of 2021. The goal was to capture drone imagery in multiple seasons for comparison. The start and end time was written down when sampling with this method to calculate the time each sampling took.

Occupancy grids were used as a reference method to capture ground measurements for percentages of warm season grass, cool season grass, legumes, soil, or weeds (Payne et al., 2021). A 5 x 5 grid was cut from a steel fence panel with 6" x 6" squares. The grid was spray painted orange for visibility in tall grass. The grid was thrown every twenty steps and the dominant species was recorded within each square. This was repeated thirty times per pasture. Species were identified by trained observers. Data were entered into Microsoft Excel. The start and end time was written down when sampling with this method to calculate the time each sampling took.

The Canopeo application within MATLAB by MathWorks® was used for Green Canopy Cover (GCC) analysis. Canopeo was developed at Oklahoma State University and is a simple, accurate, and rapid tool to analyze fractional green canopy cover from digital images (Patrignani and Ochsner, 2015). Settings used were as follows: Red/Green = 1.1; Blue/Green = 0.9; Noise reduction = 1. Noise reduction of 1 was chosen due to the high resolution necessary for detecting individual bunches of grass. A higher noise reduction setting would exclude small clusters of green pixels which is not ideal in this scenario (Patrignani and Ochsner, 2015). Blue/green and red/green tolerances are used to differentiate green pixels from the background. The levels used in this study were chosen after testing different options, including the default, to determine which ones would be most accurate for this scenario. The custom settings for color tolerances included more shades of yellow-green and pale yellow compared to the default settings. This included yellowing cool season grasses that retained less green color than others but did not include the yellow-oranges and brown-tan colors seen in dormant or senescent warm season grasses.

GCC results for the Nikon photos were averaged within each pasture using Microsoft Excel. Drone images had already been combined by orthomosaic stitching so did not require this step. Data from occupancy grid samples were grouped into three categories: warm season, cool season, and soil. The cool season species percentage from the occupancy grids was subtracted from the green canopy cover values from the same field's imagery, and then the absolute value was taken to produce a residual value (i.e., the difference in green species measured by occupancy grids and digital imagery). If occupancy grid cool season values exactly matched GCC values from imagery, the residual value would be zero.

Residual values were calculated by subtracting the occupancy grid cool season grass percentage from the imagery (Nikon or Drone) green canopy cover value (Residual = |measured GCC - reference green canopy cover). If occupancy grid cool season values matched exactly GCC values from imagery, the residual value would be zero. The absolute value of the residuals were taken since the parameter of interest was the difference between the reference method and treatment, rather than the direction of the difference.

Anova was used in RStudio to evaluate differences and interactions before proceeding with a linear mixed effects model. A linear mixed effects model was used due to the nested nature of the field within farm variables (two fields were sampled on each farm). Because of a significant interaction between sampling season and sampling method ($P < 0.05$), data were analyzed by season.

Results & Discussion: An interaction was found between sampling season and sampling method. So data are presented by season. Drone images provided the best estimate of green canopy cover in the fall, but not the spring (Fig. 1). Residual values were 12 and 22% and 14 and 5% for the drone and camera in the spring and fall, respectively. These data indicate that the drone was more consistent across seasons with an average residual of 13% and a range of 12 to 14%. Although the average for the digital-SLR was similar, the range was greater (5 to 21%).

The time required to collect data in the pastures were 12, 13, and 40 minutes for the drone, digital-SLR, and occupancy grid, respectively (Fig. 2). In addition to the time required for each method, the physical labor differed. The occupancy grid method first required training to properly identify each species, followed by the data collection in the field, and then hand-entering data points into a computer. It was by far the most strenuous method, requiring repeated standing and kneeling.

The Nikon camera was moderately labor intensive and still required a minimum of two people walking through the field. Carrying the DocaPole adds to the labor when walking over rough terrain or in hot or cold weather conditions but eliminated the requirement of kneeling repetitively. This method required the least training as species identification was not required, nor was licensing.

The least labor-intensive method was the drone. A drone can be piloted from a stationary takeoff point. If the drone remains in the pilot's visual line of sight and within legal flight requirements, no movement by the operator is necessary. Some initial mental rather than physical labor is required to program a flight path which the drone will follow. Drone flying does require a license that can be obtained in the United States from the Federal Aviation Administration by taking an exam with a $200 fee.

An important consideration in evaluating these methods is practicality. For any of these methods to be useful, they need to be affordable and doable within time and labor constraints.

The fastest methods were the drone and Nikon camera; however, they were also the costliest from an equipment standpoint. They also required the user to have or develop operational competency with these technical devices. The slowest method with the lowest equipment cost was the grid, but it required more physical labor. In addition, this method requires competency in plant identification, not commonly found in most lay persons. If this study were repeated, it would be useful to further quantify the labor by using a fitness tracker or similar device. This would also create an opportunity for labor efficiency to be compared across variable topographies, since sampling with occupancy grids or the Nikon is much easier on flat, well-groomed terrain compared to hilly, rocky terrain.

Summary & Implications: Both the Nikon and the drone show promise in their abilities to quantify cool season grasses in pastures. If this strategy is to be used for long-term documentation of botanical shifts over large areas, a drone would be more practical. If these methods are to be used at the individual farm level, more research is required to determine what level of error is acceptable in these sampling methods.

Literature Cited:

Ball, D. M., Hoveland, C. S., & Lacefield, G. D. (2007). *Southern Forages* (Vol. 4). Norcross, Georgia: International Plant Nutrition Institute.

Lacefield, G. D. (2014). Cool Season Forage Grasses. *University of Kentucky Cooperative Extension Service.*

Ohio University. (2021). 7 Pros and Cons of Drones and Unmanned Aerial Vehicles. Available at https://onlinemasters.ohio.edu/blog/the-pros-and-cons-of-unmanned-aerial-vehicles-uavs.

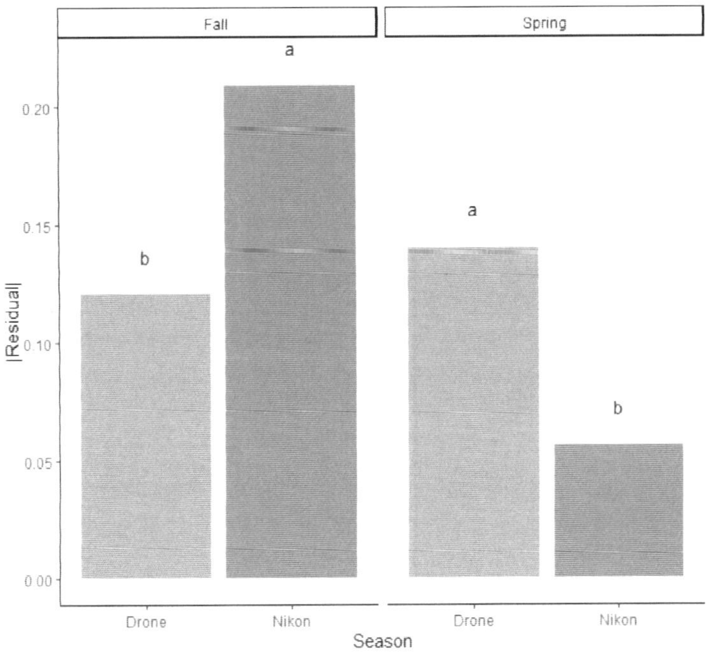

Figure 1. Residual means by method and season, calculated as the absolute value of the difference between the percentage of C3 grasses measured by occupancy grids and the green

canopy cover percentage measured from the imagery captured by each method. Bars with the same letter within the same season are not different according to an ANOVA test ($P < 0.05$).

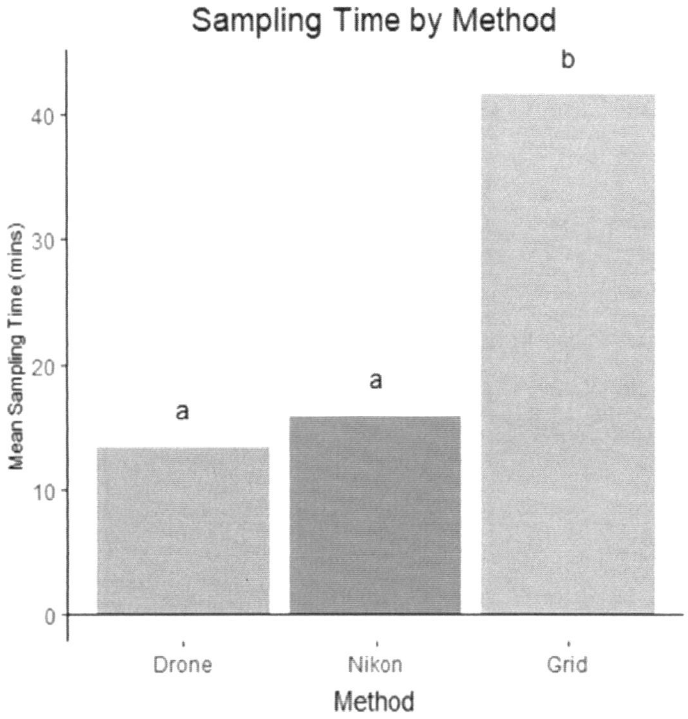

Figure 2. Sampling times for drone, digital-SLR and occupancy grid. Bars with the same letter are statistically similar according to Tukey's honestly significant difference test ($P \leq 0.05$) .

CLIMATE SMART GRASSLANDS

P.D. Keyser, C.N. Boyer, S.H. Cho, S. Jagadamma, S.M. Schexnayder, and F.R. Walker[1]

Because of the extent and importance of grasslands agriculture, even modest improvements in operational profitability, resiliency, biodiversity, greenhouse gas (GHG) reductions, and carbon sequestration could have substantial positive impacts. Indeed, grasslands are agriculture's largest and most effective carbon(C)-storage system globally as well as nationally and thus, offer a compelling opportunity for producers to engage in emerging carbon markets. In view of these goals, the University of Tennessee has been selected to receive a $30 million award in support of a large partnership (some 30 partners from Extension, industry, and trade and conservation groups) to implement a comprehensive "Climate Smart Grasslands" project across nine states at the core of the Fescue Belt. The project will engage >200 producers to implement and evaluate a suite of strategic grassland management practices and compare these to more conventional approaches. Based on a deeper and more precise understanding of on-farm C pools associated with grassland agriculture, we will identify optimum strategies for family farms to engage with the emerging C market. In addition, Extension services across these nine states will educate producers regarding innovative practices that can enhance soil C pools and/or reduce C- and other GHG-positive inputs. This information will be used to empower farmers, downstream elements of the beef value chain and allied industries to engage in carbon exchanges and foster market opportunities within the retail sector.

[1]Professor and Director, Center for Native Grasslands, (pkeyser@utk.edu); Professor, (cboyer@utk.edu); Professor, (scho9@utk.edu); Associate Professor, (sjagada1@utk.edu); Project Manager, (schexnayder@utk.edu); Professor, (frwalker@utk.edu), all with the University of Tennessee, Knoxville, TN.

CLIMATIC RESILIENCY OF KENTUCKY FORAGE SYSTEMS

Rebecca L. McCulley[1]

Pastures of the transition zone of the eastern United States are comprised of plants with both cool- and warm-season physiologies and may, thus, be well-suited to adapt to climate change. However, climate change may make pastures dominated by tall fescue (*Schedonorus arundinaceous*) and infected with the common toxic form of the fungal endophyte, *Epichloë coenophiala*, more problematic for grazing animals. To test these hypotheses, we ran a five-year study in a central Kentucky pasture where we manipulated temperature (increased 3°C over ambient conditions) and precipitation (increased 30% above mean annual) and measured a suite of forage and soil responses. As predicted, increased temperatures stimulated the abundance of warm-season grasses, particularly in the late summer and fall sample periods. However, cool-season grasses, such as tall fescue and Kentucky bluegrass, persisted in all treatments and were the dominant forages in the spring. Overall, warming stimulated forage production 19%. Forage quality responses to the climate treatments varied seasonally and across years of the study. Warming increased endophyte infection frequency within the tall fescue community and elevated toxic alkaloid concentrations, particularly in the fall. These results indicate that pastures of the eastern United States are resilient to predicted changes in climate, maintaining or enhancing overall forage production by altering forage species composition. However, for producers reliant on common toxic fescue stands, incidence and severity of fescue toxicosis may increase, suggesting a greater need for adoption of non-toxic fescue cultivars or other types of forage species moving forward.

[1]Professor and Chair, Department of Plant and Soil Sciences, University of Kentucky, Lexington, KY (rebecca.mcculley@uky.edu).

CREATING SUSTAINABLE FORAGE-ROW CROPPING SYSTEMS FOR THE SOUTHEAST WITH PERENNIAL RED AND WHITE CLOVER

Eric D. Billman[1]

Forage production in the southeastern United States is dominated by warm-season perennial or cool-season annual grass and legume species. Cotton, a crop unique to the Southeast region, is an input intensive crop that requires multiple field passes, fuel use, time spent, and labor costs incurred. Cool-season perennial forages may provide a way to create a hybrid forage-row cropping system for the region by providing a source of spring forage while acting as perennial groundcover between the rows of a cotton crop during the summer months to suppress weeds and reduce chemical and physical inputs in cotton. The objective of this study was to determine the efficacy of using perennial red and white clover as spring forage and summer groundcover in a cotton production system of the southeastern U.S. We hypothesized that a pure clover stand would result in significantly reduced weed pressure during the summer and not reduce cotton yields. By growing perennial red and white clover between the rows of a cotton crop, we observed reduced weed pressure (1 – 2 weeds ft^{-2}) with clovers compared to fallow or annual cover crop systems (10 – 15 weeds ft^{-2}). During both years of the test, red and white clover stands declined, with near total stand loss in year 2, likely due to drought stress. Cotton yields were not negatively affected by growing red and white clover with the cotton crop. These results suggest that red and white clover could be suitable for reducing input costs in southeast cotton production, provided that irrigation is available.

[1]Research Agronomist, USDA-ARS, Coastal Plains Soil, Water, and Plant Research Unit, Florence, SC (eric.billman@usda.gov).

DETERMINING THE NUTRIENT REQUIREMENTS OF ALFALFA STANDS: A COMPARISON OF SOIL AND TISSUE TESTING

W. R. Fleming, C. D. Teutsch, E. L. Ritchey, and J.G. Grove[1]

Alfalfa is a perennial forage legume that tolerates a wide range of growing conditions and is commonly produced for use in the dairy and equine industries. In the United States, alfalfa yields grew exponentially from the 1950s-1980s due to advances in synthetic pesticides and fertilizers and new varieties but have plateaued and remained stable ever since. The objective of this study was to determine the role of soil fertility in the observed yield plateau in three different regions of the United States. In 2022, 50 alfalfa stands in each of Kentucky, Oregon, and Wisconsin were sampled during the late bud to early flowering growth stage to determine their soil and tissue nutrient status. Soil testing data from Kentucky indicates that phosphorus was low in 4% and potassium was low in 33% of the sampled alfalfa stands sampled. However, tissue analysis indicated that phosphorous was considered sufficient in all sampled stands and potassium was considered limiting in 26% of sampled stands. Tissue data also indicated that magnesium and sulfur were low in 23% and 13% of sampled stands. The low sulfur status may be related to the reduction in the atmospheric deposition of sulfur. More work is needed to determine if sulfur and magnesium are truly limiting alfalfa yields in Kentucky.

Graduate Student, University of Kentucky Research and Education Center, Princeton, KY (william.fleming@uky.edu); Extension Associate Professor, University of Kentucky Research and Education Center, Princeton, KY (chris.teutsch@uky.edu); Extension Associate Professor, University of Kentucky Research and Education Center, Princeton, KY (edwin.ritchey@uky.edu); Professor of Agronomy/Soils Research & Extension, University of Kentucky Research and Education Center, Princeton, KY (jgrove@uky.edu).

DEVELOPMENT OF FORAGE-BASED EDUCATIONAL RESOURCES FOR EXTENSION AGENTS IN THE SOUTHEAST

L.S. Silva, L. Dillard, M.K. Mullenix, J. Tucker, M. Wallau[1]

The livestock industry is one of the largest agricultural activities in the Southeast. Forages are the primary feed source due to favorable climate conditions and a wide range of species adapted. Well-managed forage systems are crucial in supporting food production and environmental stewardship but require a holistic understanding of the ecosystem to balance plant, animal, soil, and environmental requirements. In this context, it is crucial to provide Extension agents and agricultural educators with easy-to-access revised educational materials and science-based training to enhance their knowledge and support technology transfer to clientele. This was a multidisciplinary collaborative effort involving several faculty and specialists from the region. A printed reference guide was developed as a textbook for a 2-day in-person training held in Alabama in April 2022. The audience (n=81) consisted of Extension agents and educators from seven states (Alabama, Florida, Georgia, Louisiana, North Carolina, South Carolina, and Tennessee). The training consisted of classroom content, hands-on training, and networking opportunities to increase experience exchange among attendees. Surveys were applied to gather feedback on the scope, educational resources, and applicability of resources in the attendees` programming activities. Regional joint efforts help address multidisciplinary training needs and revised content to support systems' productivity and profitability in the region. The major challenge continues to be funding acquisition to support the agent's attendance.

[1] Assistant Professor, Clemson University, Blackville, SC (lseveri@clemson.edu); Assistant Professor, Auburn University, Auburn, AL (dillasa@auburn.edu); Associate Associate Professor, Auburn University, Auburn, AL (mullemk@auburn.edu); Professor, University of Georgia, Tifton, GA (jjtucker@uga.edu); Assistant Professor, University of Florida, Gainesville, FL (mwallau@ufl.edu). This effort was supported by the SARE PDP award# SPDP21-04.

EFFECTS OF CATTLE REMOVAL DATE ON SUSTAINABILITY OF COOL-SEASON ANNUAL GRAZED COVER CROP SYSTEMS

M. B. Henson[1,] R. C. Carrell[1], S. L. Dillard[1, 2], M. K. Mullenix[1],
A. V. Gamble[2], R. B. Muntifering[1]

Grazing of cool-season cover crops has been shown to be a viable tool in extending the grazing season while mitigating environmental risks associated with row crop farming systems. Our objective was to evaluate the effects of cattle removal date on forage yield, forage quality, and steer performance when grazing cool-season cover crops. A forage mix consisting of 'Cosaque' oats (*Avena strigose)*, 'FL401' cereal rye (*Secale cereal)*, 'AU Sunrise' crimson clover (*Trifolium incarnatum)*, and 'T-raptor' brassica (*Brassica napus* × *B. rapa*) was established in twelve, 0.62 ha paddocks, which were randomly allocated to one of four cattle removal treatments: no grazing (CON); February removal (FEB); March removal (MAR); or April removal (APR). Three tester steers were randomly allotted, based on body weight (BW), to each treatments except CON. Steers were weighed every 30 d to determine final BW and average daily gain (ADG). At inception and every two weeks, forage was harvested to analyze average forage yield, neutral detergent fiber (NDF), acid detergent fiber (ADF), and crude protein (CP). Significant differences in forage yield were detected between CON and FEB (3,913 vs. 2,261 kg/ha), CON and MAR (3,913 vs. 1,790 kg/ha), and CON and APR (3,913 vs. 1,810 kg/ha). There were no differences in ADG (1.2 kg/day), NDF (50.94%), or ADF (30.34%) among treatments. Results indicate that grazing of cool-season annual cover crops will reduce forage yield, but removal date of cattle will not further affect forage yield, quality, or individual cattle performance, reducing winter supplementation for cattle by increasing grazing days.

[1]Department of Animal Science, Auburn University, Auburn, AL; [2]Department of Crop, Soil, and Environmental Sciences, Auburn University, Auburn, AL.

EFFECTS OF NITROGEN RATE AND CRIMSON CLOVER ADDITION ON TRITICALE PERFORMANCE

D. D. Harmon[1] and E. G. Henderson[2]

Opportunity exists for cool-season annual forages to extend the grazing season and provide high quality feed for beef cattle in Appalachia. With recent increases in fertilizer prices, producers have renewed interest in the nitrogen fixation properties of clovers and precision nutrient applications to maximize return on investments. The objective of this study is to determine forage mass and nutritive value of triticale (*Triticosecale Wittm.*) seeded with and without crimson clover (CC; *Trifolium incarnatum*) and fertilized with 6 nitrogen (N) rates. In the fall of 2021, triticale ('Surge') was planted into a prepared seedbed in a monoculture at 105 lb/acre pure live seed (PLS) or in a mixed sward at 75 lb/acre PLS with 15 lb/acre PLS of crimson clover ('Dixie Reseeding'). At planting, all plots received 40 lb N/acre and an additional 0, 30, 60, 90, 120 or 150 lb N/acre upon green up in spring. Plots were harvested at flowering stage, subsampled, and analyzed for nutritive value. Forage mass was greater for triticale planted with CC than planted without (7,722 and 7,245 lb/acre), but did not differ among N rate. Clover inclusion had no effect on crude protein (CP) content, but CP increased linearly with increasing N rate. Triticale interseeded with CC had greater total digestible nutrients (TDN) than triticale seeded without CC (57.9 and 56.30% TDN), but N rate did not differ. Ash concentration increased linearly with N rate and was greater for triticale planted without CC than with CC (7.56 vs 7.11% DM). The results indicate that adding a legume may be an effective alternative to N fertilizer in increasing forage yields and nutritive value in triticale forage crops.

[1] Livestock Extension Specialist, N.C. State University Dept. of Animal Science, Waynesville, NC (ddharmon@ncsu.edu); [2]Livestock Marketing Consultant, Virginia Department of Agriculture and Consumer Services, Wytheville, VA (ethan.henderson@vdacs.virginia.gov).

EFFECTS OF SPRING AND FALL STOCKPILE GRAZING MANAGEMENT SYSTEM ON PERFORMANCE OF FALL-CALVING BEEF COWS, PRE-WEANING CALF PERFORMANCE, AND FORAGE CHARACTERISTICS

K. P. Ewing, L. T. Neira, W. T. Meteer, T. L. Steckler, and D. W. Shike[1]

The objective of this study was to compare the effects of tall fescue grazing systems either utilizing rotational grazing only or rotational grazing and spring and fall stockpiling on beef cow body condition score and reproductive performance, calf pre-weaning performance, and forage characteristics. Over two years, multiparous, fall-calving cows (n = 360/year) were randomly assigned to one of two treatments; 1) rotational grazing and spring and fall stockpile grazing (**SP**) and 2) rotational grazing only (**CON**). The CON pastures were divided into eight paddocks. Twenty-five percent of SP pastures were stockpiled from treatment initiation with remaining 75% rotationally grazed. An additional 50% was stockpiled starting early August. Strip-grazing was utilized when grazing stockpiled forage in SP pastures. Grazing days were increased by 24 days for SP compared to CON. Hay fed per cow was reduced in SP compared to CON (130 vs. 765 lb, respectively). Cow body weight (BW) did not differ at trial end however, body condition score was greater for CON than SP (5.3 vs. 5.1, respectively). No differences were observed in AI or overall pregnancy rate between treatments. Calf birth and weaning BW were similar between treatments. Utilization of rotational grazing and spring and fall stockpiling system had little impact on cow and calf performance but increased grazing days and reduced hay fed to cows. This novel grazing strategy can be implemented in tall fescue systems to extend the grazing season, reduce stored feed use, and potentially reduce winter feed costs.

[1]University of Illinois at Urbana-Champaign, Animal Sciences Laboratory, Urbana, IL

ESTABLISHING ALFALFA THROUGH SUNFLOWER INTERCROPPING

M. Tabert[1], M.T. Berti[1], and J.V. Anderson[2]

Intercropping is a way to better utilize the growing season and available resources more profitably. The integration of alfalfa (*Medicago sativa* L.) into crop rotations may also lead to a reduction of synthetic nitrogen use and losses by leaching or run-off. This research evaluated the possibility of intercropping alfalfa with sunflower (*Helianthus annuus* L.) in the northern Great Plains using a randomized complete block design to establish alfalfa for the subsequent crop year. This approach could increase alfalfa forage yield and nutritive value in the second year while providing a sunflower crop during the normally low-producing alfalfa establishment year. Treatments were alfalfa alone, sunflower alone at 30- and 60-inch row spacing, and sunflower intercropped with alfalfa at 30- and 60-inch row spacing. Sunflower achene yield in the first year was measured and no significant differences among treatments were determined in 2021. Alfalfa forage yield and nutritive value was also measured and calculated in 2021, along with measurements of soil gravimetric water and light interception. There was an interaction between treatments and date for light interception, mainly because alfalfa alone had much lower light interception than all treatments with sunflower. Gravimetric water was lower in soil where alfalfa and sunflower were intercropped. In-row alfalfa dry biomass and quality were not significantly different for the 30- or 60-inch sunflower row spacings.

[1]M.S. Student, Department of Plant Sciences, North Dakota State University, Fargo, ND (mikayla.miller@ndsu.edu); [1]Professor, Forages, Cover Crops, and Biomass Production, North Dakota State University, Fargo, ND (marisol.berti@ndsu.edu); [2]Research Chemist USDA-ARS Edward T. Schafer Agricultural Research Center, Fargo, ND (james.v.anderson@usda.gov).

EVALUATING RESILIENCY TRAITS IN AGRICULTURAL PRODUCTION SYSTEMS

S. Gruss[1], K.A. Cassida[1], M. Berti[2], J. Guretzky[3], E. Van Santen[4], R. Oakes[5], A. Islam[6]

The increasing frequency of extreme weather events, inflation leading to higher input costs, and increasing world population places more pressure on our agricultural production systems to maintain while meeting production goals during these harsh influxes. Current agricultural systems are dominated by annual crop monocultures, that be heavily influenced by the above factors. We propose that the integration of perennial crops with an increase in circular practices and diversity of crops can build a more resilient and sustainable system. Our project aims to understand different resiliency factors by comparing two systems: a local dominant system (LDS) and a diverse, perennial, circular system (DPCS) through 50-farm pairs across the United States and ecoregions. Pairs will have differing production systems (LDS vs. DPCS) and be geographically close with similar soil types. Each farm will be examined for soil health, agronomic information, and socioeconomic factors across three years. Data will be utilized to understand each system's decision-making processes, resiliency, and evaluate obstacles inhibiting the integration of more diverse, circular, and perennial practices. Understanding the resiliency factors for each farm system can benefit in creating a more dynamic system, create opportunities to influence policy to help overcome potential socioeconomic barriers, and lead to the dissemination of information to producers by building a network of farmers across the US.

[1]Research Assoc. and Assoc. Professor, Dept. of Plant, Soil, and Microbial Sciences, Michigan State Univ., East Lansing, MI (grussshe@msu.edu; cassida@msu.edu), [2]Professor, Dept. of Plant Sciences, North Dakota State University, Fargo, ND (marisol.berti@ndsu.edu), [3]Associate Professor, Dept. of Agronomy and Horticulture, University of Nebraska, Lincoln, NE (jguretzky2@unl.edu), [4]Professor, Agronomy Dept., University of Florida, (evsanten@ufl.edu), [5]Associate Professor, Dept. of Plant Sciences, University of Tennessee, Knoxville, TN (rnave@tennessee.edu), [6]Professor, University of Wyoming, Dept. of Plant Sciences, Laramie, WY (mislam@uwyo.edu).

EFFECTS OF RYE OR TRITICALE WITH OR WITHOUT CRIMSON CLOVER ON FORAGE YIELD AND QUALITY, AND GESTATING BEEF FEMALE PERFORMANCE

K.P. Ewing and D.W. Shike[1]

The objective of this study was to evaluate the effects of cereal rye or triticale with or without crimson clover on forage yield and quality, and gestating beef female performance. In experiment 1, four cover crops were randomly assigned to the treatments cereal rye or triticale with crimson clover or without crimson clover. Plots were seeded mid-September (cereal rye or triticale: 90 lb/acre; crimson clover inclusion: 70 lb/acre cereal and 8 lb/acre legume). Forage yield was greater with inclusion of crimson clover than without (2,862 vs. 2,368 lb/acre, respectively). Crimson clover inclusion reduced dry matter (DM) but increased CP (9.6% vs. 8.6%, respectively). Cereal rye had greater DM, neutral detergent fiber and acid detergent fiber compared to triticale. In experiment 2, spring-calving cows and heifers were randomly assigned to one of five treatments in a 42-day feeding trial. Treatments were similar to experiment one with the addition of alfalfa hay (CON). All treatments were limit fed (25 lb/hd/d DM basis). Final body weight (BW), BW change, body condition score (BCS), and BCS change were greater in cover crop treatments compared to CON. Body weight change did not differ between cereal type or with inclusion of crimson clover. Crimson clover inclusion increased BCS change. In summary, inclusion of crimson clover increased forage yield, CP, and female BCS. Although cereal rye cover crops had decreased quality prior to ensiling, there were no differences in cow performance compared to triticale. Substituting cover crops for alfalfa hay resulted in increased cow BW and BCS, suggesting the cover crops evaluated have greater feed value than alfalfa hay.

[1]University of Illinois at Urbana-Champaign, Animal Sciences Laboratory, Urbana, IL

FALL NITROGEN IMPROVED SPRING YIELD FOR TALL FESCUE HAYFIELDS

B.L. Hendrix, C.D. Teutsch, J.R. Duckworth[1]

Late season nitrogen fertilization is a practice that has long been implemented in turfgrass production. This results in an extended green period in the fall without stimulating excessive shoot growth as well as earlier green up in the spring. The longer green period results in higher levels of carbohydrate reserves and enhanced root growth compared with a spring-summer fertilization program. However, limited research has considered the impact of late fall nitrogen applications on grass growth in hayfields and pastures. The objective of this study was to evaluate the effects of late fall nitrogen fertilization on dry matter production of an established tall fescue stand. This experiment was conducted at the UK Research and Extension Center located in Princeton, KY in 2020 and 2021. The experimental design was a random complete block with four replications. In early December 2020 and 2021, 0, 30, 60 and 90 lb N/acre was applied as ammonium nitrate, ammonium sulfate, and Anuvia SYMTRX 20S. An additional 80 lb N/acre was applied as urea to all plots the following spring. Plots were harvested in May using a self-propelled forage harvester equipped with load cells. Fall applied nitrogen resulted in a linear increase in dry matter yield of 22.3 and 18.9 lb DM/acre in 2021 and 2022, respectively. Nitrogen source had no impact on dry matter yield ($P > 0.05$). Both nitrogen rate and source had limited impacts on CP, ADF, and NDF. Additionally, plots receiving fall applied nitrogen initiated growth earlier in the spring than those which only received a spring application.

[1]Senior Research Technician, University of Kentucky Research and Education Center, Princeton, KY (brittanyhendrix@uky.edu); Extension Associate Professor, University of Kentucky Research and Education Center, Princeton, KY (chris.teutsch@uky.edu); Agronomist I, University of Kentucky Research and Education Center, Princeton, KY (joshua.duckworth@uky.edu)

FEED SUPPLEMENTATION WITH NATURAL RED CLOVER PRODUCT, BIOCHANIN A, DECREASES TRACE GAS EMISSIONS FROM SOIL-APPLIED LIVESTOCK WASTE

A.A. Jacobs[1], M.D. Flythe[2], D.G. Ely[3], L. Muñoz[3], J. May[2], J.A. Nelson[1], V. Stanton[4], K. Pham[1], and R.L. McCulley[1]

Feeding natural plant metabolites is a new strategy to increase livestock productivity. Feeding cattle and sheep the isoflavone biochanin A (BCA), a plant secondary metabolite naturally produced by red clover, improves livestock weight gain, but little is known about corresponding environmental impacts in pastures. Livestock production currently emits the greatest amounts of greenhouse gases within the agricultural sector, where urine excreta patches in pastures are hotspots for soil trace gas emissions. To determine BCA supplementation effects on urine patch trace gas emissions, urine collected from lambs that were fed no BCA, 0.45, or 0.90 g/day BCA was applied to soil in a laboratory incubation and trace gases were measured over a 60-day period. Feeding lambs BCA reduced total ammonia volatilization by 35%. Nitrous oxide and methane emissions were reduced by 78% and 5%, respectively, on days when gas concentrations significantly differed. A separate 13-day incubation using urine from a single lamb fed no BCA and spiked with nine levels of BCA confirmed that BCA reduced soil emissions of methane, nitrous oxide, and ammonia. Reductions in trace gases did not correspond to changes in soil microbial structure, indicating BCA altered other aspects of urine patch dynamics, such as soil enzyme activity. Overall, results of this study show livestock BCA supplementation may mitigate greenhouse gas emissions in the livestock industry and support sustainable production goals.

[1]Dept. of Plant and Soil Sciences, University of Kentucky, Lexington, KY; [2]USDA-ARS Forage Animal Production Research Unit, Lexington, KY; [3]Dept. of Animal and Food Sciences, University of Kentucky, Lexington, KY; [4]Applied Statistics Laboratory, University of Kentucky, Lexington, KY. Corresponding author: alayna.jacobs@uky.edu

DETERMINING THE NUTRIENT REQUIREMENTS OF ALFALFA STANDS: A COMPARISON OF SOIL AND TISSUE TESTING

Fleming, W. R., Teutsch, C. D., Ritchey, E. L., Grove, J. H.[1]

Abstract

Alfalfa (*Medicago sativa* L.) is a perennial forage legume and is sometimes referred to as "The Queen of Forages" due to its high nutritional value. It is commonly produced as hay for use in the dairy and equine industries. Alfalfa yield increased exponentially from the 1950s to the1980s due to new breeding and management innovations; however, during the 1980s yields plateaued at around 3.3 tons/ac and remain there today. The overarching goal of this study was to determine the role of soil fertility in the observed yield plateau. Fifty alfalfa stands in Kentucky were sampled to gather information about their macro and micronutrient status. Samples were taken when the stand reached the late bud to early flower stage of maturity from a 20 x 20 ft^2 area that was representative of the entire stand. Soil samples were collected at 4 and 6 in and were sent to the University of Kentucky soil testing lab for analysis. The top 6 in of 30 stems were dried, ground, and sent to Kansas State University for tissue analysis. In 2022, soil and tissue data from the alfalfa stands sampled indicated that potassium, sulfur, and magnesium were the nutrients most commonly reported to be below established sufficiency ranges. In addition, pH was low in just under half of the sampled stands. Tissue analysis indicated that soil testing overestimated potassium deficiencies in alfalfa stands in Kentucky. Even so, potassium levels were below the sufficiency range in nearly one-fourth of the alfalfa stands sampled.

Introduction: Alfalfa (*Medicago sativa* L.), also known as the "Queen of Forages" due to its high forage quality, is a perennial forage legume and the fourth most widely grown crop in the United States. The most common use for alfalfa is as hay and haylage for dairy cattle but it is also used as feed for equine and other livestock (USDA-ARS, 2020). Alfalfa yields in the United States grew exponentially from the 1950s to the 1980s; however, in the 1980s alfalfa yields plateaued where they remain at approximately 3.3 tons/ac today (USDA-NASS, 2022). Increases in yields during that time are accredited to the development of new cultivars with resistance to multiple pests, higher yield potential, and improved management practices (Barns et al., 1988). The overarching objective of this study was to determine the role of soil fertility in the observed yield plateau in three different regions of the United States. This paper will only present data from Kentucky for the 2022 growing season.

[1] Graduate Student, University of Kentucky Research and Education Center, Princeton, KY (william.fleming@uky.edu); Extension Associate Professor, University of Kentucky Research and Education Center, Princeton, KY (chris.teutsch@uky.edu); Extension Associate Professor, University of Kentucky Research and Education Center, Princeton, KY (edwin.ritchey@uky.edu); Professor of Agronomy/Soils Research & Extension, University of Kentucky Research and Education Center, Princeton, KY (jgrove@uky.edu).

Materials and Methods: Samples were collected from 53 alfalfa stands across Kentucky. Stands were chosen based on the ability to obtain the management history and the producer's willingness to participate in the study. Stands were from 1-5 years of age and sampling was conducted when plants were between the late bud to the early flower stage of maturity. Samples were collected from a 20 x 20 ft^2 area that was representative of the entire stand.

Composite soil samples were collected from within the 20 x 20 ft^2 area at depths of 4 and 6 inches and dried at 131°F before being sent to the University of Kentucky soil testing laboratory. Analysis of P, K, S, Ca, Mg, Mn, Cu, Zn, and Fe was performed using the Mehlich 3 extraction method. Boron analysis was performed using hot water extraction (2020-2021 Lime and Fertilizer Recommendations). Soil water pH and Sikora buffer pH were also determined. Tissue nutrient status was quantified by collecting the top 6 in of 30 random stems from each 20 x 20 ft^2 area and drying them for 3 days at 131°F. Dried samples were ground to pass 0.08 and 0.04 in screens using Wiley (Thomas Scientific, Swedesboro, NJ) and Cyclone (Udy Corp., Fort Collins, Co) sample mills, respectively. Ground samples were sent to Kansas State University for N, P, K S, Ca, Mg, Mn, Cu, Zn, and Fe analysis.

Results and Discussion: Approximately 40% of stands had a pH level below the ideal range (Table 1). Low pH has been shown to cause declines in nitrogen fixation and inhibition of nodulation in legumes. Low pH also impacts the availability of other nutrients in the soil (Lanyon and Griffith, 1988). Rice and coauthors (1977) found that yield declines alfalfa were reported at pH levels below 6.0. Further, Walworth and Sumner (1990) reported that liming increased both dry matter production and persistence of alfalfa.

Soil tests averaged over the 4 and 6 in sampling depths indicate that phosphorus was low in 5% of sampled stands while potassium was low in 33% (Figure 1). However, the plant tissue analysis indicated that phosphorus was sufficient in all samples and potassium was below the sufficiency range in 26% of sampled stands (Table 2). This discrepancy was likely due to some soil samples falling into the upper portion of the low range where yield may not have been limited.

Low phosphorus or potassium can inhibit alfalfa yields. Jungers et al. (2019) found that potassium application increased yields in four to five cuttings per year management system; however, they found it lowered forage quality. It has been reported that applications of phosphorus and potassium or phosphorus alone after the final cutting increased the yields in the first cutting in the spring. However, potassium did not report a yield increase, indicating that phosphorus levels may impact winter hardiness (Berg et al., 2021). Walworth and Sumner (1990) found that potassium applied in the spring increased yields, however in two out of the three years this only was observed when magnesium was applied with it. They also found that magnesium levels in the soil and tissue were suppressed when potassium was applied. Magnesium applied on its own decreased dry matter production (Walworth and Sumner, 1990). They concluded that these two nutrients are in competition with each other. This offers a potential explanation for the low magnesium levels reported in Table 2.

Tissue analysis indicated sulfur was below the sufficiency range in 13 percent of the stands that were sampled (Table 2). Currently, the University of Kentucky does not have a soil test recommendation for sulfur in alfalfa production. Low sulfur could hinder symbiotic nitrogen fixation. Pumphrey and Moore (1965) reported yield increases and increased shoot nitrogen content when they applied sulfur to alfalfa growing under low sulfur conditions. Reductions in fossil fuel use have caused sulfur deposition to decrease across the United States resulting in lower sulfur concentrations in soils. The United States Environmental Protection Agency reported that sulfur deposition across Kentucky reduced from 9 to 16 lb. S/ac to 0 to 5 lb. S/ac over the last 20 years.

In conclusion, soil and tissue analyses indicate that alfalfa yields may be limited in some cases by soil fertility. Soil pH was below the optimal range in more than 40% of the fields sampled. This may result in decreased nitrogen fixation and nutrient availability. Potassium was reported low in approximately one-quarter of the sampled stands according to tissue analysis. This was not unexpected since hay production removes large quantities of potash. Sulfur and magnesium were reported low in 13 and 23 percent of stands, respectively. More work is needed to better understand if these two nutrients are truly limiting alfalfa yield in Kentucky.

Literature Cited:

2020-2021 Lime and Nutrient Recommendations, AGR-1. 2020. University of Kentucky Cooperative Extension Service, Lexington.

Barnes, D. K., Goplen, B. P., & Baylor, J. E. 1988. Highlights in the USA and Canada. *In* Hanson, A. A., Barnes, D. K., & Hill, R. R. *Alfalfa and Alfalfa Improvement.* American Society of Agronomy, Inc. (1st ed., pp. 2-22).

Berg, W. K., Brouder, S. M., Cunningham, S. M., & Volenec, J. J. 2021. Potassium and Phosphorus Fertilizer Impacts on Alfalfa Taproot Carbon and Nitrogen Reserve Accumulation and Use During Fall Acclimation and Initial Growth in Spring. *Frontiers in plant science, 12*, 715936. https://doi.org/10.3389/fpls.2021.715936

Jungers, J. M., Kaiser, D. E., Lamb, J. S, Lamb, J. A., Noland, R. L., Samac, D.A, Wells, M. S., and Sheaffer, C. C. 2019. Potassium fertilization affects alfalfa forage yield, nutritive value, root traits, and persistence. *Agronomy Journal, 111*(6) 2843-2852. https://doi-org.ezproxy.uky.edu/10.2134/agronj2019.01.0011.

Lanyon, L E. and Griffith, W. K. 1988. Nitrogen and fertilizer use. *In* Hanson, A. A., Barnes, D. K., & Hill, R. R. *Alfalfa and Alfalfa Improvement.* American Society of Agronomy, Inc. (1st ed., pp. 2-22).

Pumphrey, F. V., & Moore, D. P. 1965. Sulfur and nitrogen content of alfalfa herbage during growth. *Agronomy Journal, 57*(3), 237-139. https://doi.org/10.2134/agronj1965.00021962005700030001x

Rice, W. A., Penney, D. C., & Nyborg, M. 1977. Effects of soil acidity on rhizobia numbers, nodulation and nitrogen fixation by alfalfa and red clover. *Can. J. Soil Sci. 57.* 197-203. https://doi.org/10.4141/cjss77-024

United States Department of Agriculture: Agricultural Research Service. (Nov., 2020). *Alfalfa.* Alfalfa : USDA ARS.

United States Department of Agriculture-National Agricultural Statistics Service. 2022. Alfalfa Hay Yield from 1919 to 2020.
Walworth, J. L. & Sumner, M. E. 1990. Alfalfa response to lime, phosphorus, potassium, magnesium, and molybdenum on acid ultisols. *Field Research, 24*. 167-172. https://doi.org/10.1007/BF01073585.

Soil pH	Proportion of Stands
	%
High (>7.0)	23.00
Ideal (6.5 to 7.0)	36.00
Low (6.0 to 6.4)	30.00
Very Low (<6.0)	11.00

Table 1. Proportion of sampled stands at each pH range.

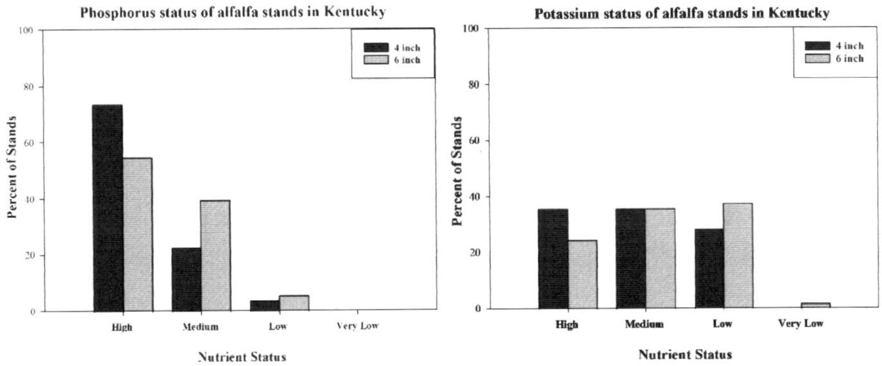

Figure 1. Percent of soil samples falling into the high, medium, low, and very low range for soil test phosphorus and potassium at 4 and 6-inch soil sampling depths.

Nutrient Status	P	K	S	Mg
	------------------% of fields sampled---------------			
High	0	0	0	0
Sufficient	100	74	87	77
Low	0	26	13	23

Table 2. Percent of sampled alfalfa fields falling into the high, sufficient, and low ranges for P, K, S, and Mg as indicated by plant tissue testing.

FORAGE BIOMASS, GRAIN YIELD, AND ECONOMIC EVALUATION OF FOUR WHEAT VARIETIES IN THE SOUTHEASTERN U.S.

M.H. West, M.K. Mullenix, A.N. Rabinowitz, W.B. Smith, and S.L. Dillard[1]

Dual-purpose wheat (*Triticum aestivum* L.) systems increase farm sustainability by diversifying on-farm income. This study aimed to evaluate pre- and post-grazing forage biomass of four winter wheat varieties managed under a dual-purpose grazing and grain production system. The wheat varieties evaluated were generic feed-type wheat (unknown variety blend, Feed), seed-type wheat 'GA Gore' (Seed) and two forage-type varieties, 'AGS 2024' (AGS) and 'Pioneer 26R41' (Pioneer). The experiment was a randomized complete block design ($n = 4$) that utilized three grazing frequencies: an ungrazed control (NG), low frequency (LF), and high frequency (HF) grazing schedule. Forage biomass was determined using three 0.1m^2-quadrats per plot before (Pre-G) and after (Post-G) each grazing event. Data were analyzed using PROC GLIMMIX of SAS (SAS Inst., Cary, NC) with differences were declared at $P < 0.05$. Prior to grazing, AGS forage biomass (2,360 lb/a) was greater ($P \leq 0.03$) than all other varieties. Across grazing frequencies, Pioneer had greater final grain yield (57.6 bu/a; $P < 0.01$) with Seed having the least (20.2 bu/a; $P < 0.01$). These results indicate that common Southeastern wheat varieties can be successfully utilized in a dual-purpose management system.

[1]Graduate Research Assistant, (mhw0030@auburn.edu); Extension Specialist (mullemk@auburn.edu); Assistant Professor (wbs0001@auburn.edu); Extension Specialist (adam.rabinowitz@auburn.edu; Extension Specialist (dillasa@auburn.edu); Auburn University, Auburn, AL.

HEMP AS A FORAGE

Swarup Podder[1], John Fike[1], and Bain Wilson[2]

Hemp (*Cannabis sativa* L.) has been a value to humans for much of our history given the plant's broad adaptation and multiple uses. Farmers are interested in hemp as a forage and feedstuff given that the plant's rapid summer production and unique nutritional characteristics (particularly seed fatty acid profile). Pharmacological properties (e.g. cannabinoids in leaf and bud tissues) may also benefit animal health, productivity and product quality under different regulatory framework. Plenty of leaf biomass and post-extracted flower biomass is available from CBD hemp processing and could be a potential co-product forage source. We analyzed various hemp forage samples to evaluate feed value. The samples included post-processed leaves from Virginia Tech CBD, post-extracted flowers (collected from farmers in 2019), and eight cultivars (Grandi, Joey, Anka, Bialo, 452, 2730, Jinma, Elletta) harvested at seven weeks in Blacksburg, VA. Two cultivars (Grandi and Joey) were also managed solely as forage crops and sampled at different growth stages in 2021 and 2022. Leaves contained 20.6% crude protein (CP), post extracted flowers contained 25.6% CP, dual-purpose (grain+fiber) cultivars contained 15.9 to 20.1% CP, and fiber cultivars (vegetative stage) contained 24 to 27.3% CP. Grandi and Joey contained 20 to 31% CP, 24 to 44% NDF and 22 to 38% ADF, and 4 to 9% lignin. These results suggest that hemp has potential as a forage crop, but research efforts will need to address field establishment and production management, harvest timing (to optimize tonnage and forage quality), along with studies of animal intake and performance.

[1]Department of Crop and Soil Environmental Sciences, Virginia Tech, Blacksburg, VA-24061;
[2]Department of Animal Sciences, Ohio State University, Columbus, OH 43210.

FALL NITROGREN IMPROVED SPRING YIELD FOR TALL FESCUE HAYFIELDS

B.L. Hendrix, C.D. Teutsch, J.R. Duckworth[1]

Abstract

Late season nitrogen fertilization is a practice that has long been implemented in turfgrass production. This practice involves applying nitrogen from September through December and results in an extended green period in the fall without stimulating excessive shoot growth. The longer green period results in higher levels of carbohydrate reserves and enhanced root growth compared with a spring-summer fertilization program. In addition to these benefits, sods fertilized in late fall have been shown to green up two to six weeks earlier in the spring. Limited research has considered the impact of late fall nitrogen applications on grass growth in hayfields and pastures. The objective of this study was to evaluate the impact of late fall nitrogen fertilization on dry matter production of an established tall fescue stand. In 2020 and 2021, the study was conducted at the UK Research and Extension Center located in Princeton, KY. The experimental design was a random complete block with four replications. In early December 2020 and 2021, 0, 30, 60 and 90 lb N/A was applied as ammonium nitrate, ammonium sulfate, and Anuvia SYMTRX 20S. An additional 80 lb N/A was applied as urea to all plots the following spring. Plots were harvested on May 13th 2021 and May 24th 2022 using a self-propelled sickle bar type forage harvester equipped with load cells. Averaged across sources, first harvest yields ranged from 2617 to 4580 and 3160 to 4796 lb DM/A in 2021 and 2022, respectively. Fall applied nitrogen resulted in a linear increase in dry matter yield of 22.3 and 18.9 lb DM/A in 2021 and 2022, respectively. Nitrogen sources had no impact on dry matter yield ($P > 0.05$). Plots receiving fall applied nitrogen also initiated growth earlier in the spring.

<u>Introduction:</u> Tall fescue (*Schedonorus arundinaceus* (Schreb.) Dumort., nom. cons.) is the most important cool-season grass species in the United States (Hoveland, 2009), providing pasture for more than 10 million brood cows (USDA-NASS). This grass is known to be long-lived and robust, largely due to its mutualistic relationship with an endophytic fungus (Hoveland, 2009). Late season nitrogen fertilization is a practice that has long been implemented in turfgrass production (Steet et al. 2015). This practice involves applying nitrogen from September through December, and results in an extended green period in the fall without stimulating excessive shoot growth (Street et al., 2015). The longer green period results in higher levels of carbohydrate reserves and enhanced root growth compared with a spring-summer fertilization program. In addition to these benefits, sods fertilized in late fall have been shown to green up two to six

[1] Senior Research Technician, University of Kentucky Research and Education Center, Princeton, KY (brittanyhendrix@uky.edu); Extension Associate Professor, University of Kentucky Research and Education Center, Princeton, KY (chris.teutsch@uky.edu); Agronomist I, University of Kentucky Research and Education Center, Princeton, KY (joshua.duckworth@uky.edu)

weeks earlier in the spring (Street et al., 2015). Limited research has considered the effect of late fall nitrogen applications on grass growth in hayfields and pastures. The objective of this study was to evaluate the impact of late fall nitrogen fertilization on dry matter production of an established tall fescue stand.

Materials and Methods: An experiment looking into the potential benefits of fertilizing pastures and hayfields in the fall was conducted at the University of Kentucky Research and Education Center located in Princeton, KY in 2020 and 2021. A randomized complete block design with four replications was set up for this experiment in an established tall fescue hayfield. Plots were 10 x 20 ft, with 5 ft alleys. Four different application rates of three different sources of nitrogen fertilizer were applied in early December. Ammonium nitrate (34-0-0), ammonium sulfate (21-0-0-24S), and Anuvia SYMTRX 20S (16-1-0-20S) were spread by hand at rates of 0, 30, 60 and 90 lb N/A. Beginning in March of the following spring, compressed sward heights were collected using a rising plate meter. Three measurements were taken down the center of each plot and the values recorded. This process was continued weekly until the first harvest in May. A blanket spring application of urea (46-0-0) was applied at 80 lb N/A to all plots in late March 2021 and early April 2022. A 5 ft wide strip was harvested down the center of each plot using a self-propelled sickle bar type forage harvester equipped with load cells. Yield weights from each plot were recorded and later used to determine forage yield per acre. Subsamples were collected from each plot for dry matter and nutritive value determinations. Subsamples were dried, ground, and then scanned through a near infrared spectrophotometer (McIntosh, 2022). Data were analyzed using the polynomial regression (SigmaPlot 15.0) and the general linear model procedure (SAS, Cary, NC).

Results and Discussion: There was a significant year by treatment interaction for source, so data are being presented by year. Within years, no significant nitrogen rate by nitrogen source interaction was observed, so the main effects of nitrogen rate and source are presented. Averaged across sources, first harvest yields in 2021 ranged from 2617 to 4580 lb DM/A for 0 and 90 lb N/A, respectively. For 2022, first harvest yields ranged from 3160 to 4796 lb DM/A for 0 and 90 lb N/A, respectively. Fall applied nitrogen resulted in a linear increase in dry matter yield of 22.3 and 18.9 lb DM/lb fall applied N in 2021 and 2022, respectively (Figure 1). Nitrogen source had no impact on dry matter yield (Figure 2). Nutritive value data is summarized for the main effects of nitrogen source and rate in Tables 1 and 2. Both nitrogen rate and source had limited impacts on CP, ADF, and NDF.

This study indicates that fall applications of nitrogen have the potential to increase first harvest yield in forage systems. Fall nitrogen can encourage tillering in tall fescue plants, leading to thicker and more resilient stands that have a jump-start for spring growth. This experiment should be repeated in other locations to generate a more robust dataset that could be regionally applied. Additionally, expanding the treatment protocol to include plots that only receive fall or spring nitrogen could help us better understand the effects of application timing and the possible benefits of fall nitrogen over traditional spring applications.

Literature Cited:

Ball, D. M., Hoveland, C. S., & Lacefield, G. D. 2015. Southern Forages: Modern Concepts for Forage Crop Management. International Plant Nutrition Institute (IPNI).

Hoveland, C. S. 2009. Origin and history. p. 3-10. *In* H.A. Fribourg, et al. (eds) Tall Fescue for the Twenty-first Century. Agron. Monogr. No. 53. ASA-CSSA-SSSA. Madison, WI.

Lacefield, G., Ditsch, D., Smith, S.R., Henning, J., & Johnson, K. 2000. AGR-175: Forage Identification and Use Guide. Retrieved from: http://www2.ca.uky.edu/agcomm/pubs/AGR/AGR175/AGR175.pdf

McIntosh, D., B. J. Anderson-Husmoen, R. Kern-Lunbery, P. Goldblatt, R. Lemus, T. Griggs, L. Bauman, S. Boone, G. Shewmaker, and C. Teutsch. 2022. Guidelines for Optimal Use of NIRSC Forage and Feed Calibrations in Membership Laboratories (Second Edition). The University of Tennessee Press, Knoxville, TN.

Ritchey, E., McGrath, J. 2020. AGR-1: 20-21 Lime and Nutrient Recommendations. University of Kentucky Cooperative Extension Services. Retrieved from: http://www2.ca.uky.edu/agcomm/pubs/agr/agr1/agr1.pdf

Street, J.R., P.J. Sherratt, and K.T. Danneberger. 2015. The benefits of late season fertilization. OSU Spots Turf Management. Available at https://u.osu.edu/athleticfieldmanagement/2015/11/24/the-benefits-of-late-season-fertilizer/

U.S. Department of Agriculture. *Quick Stats.* Washington, DC: U.S. Department of Agriculture, National Agricultural Statistics Service. Available online at https://quickstats.nass.usda.gov/

Table 1. Crude protein (CP), Acid detergent fiber (ADF) and Neutral detergent fiber (NDF) values (%) for each nitrogen source averaged across all rates in both trials. Least significant difference (LSD) is given unless the values were not significant (NS).

			Nitrogen Source			
			Anuvia SYMTRX 20S	Ammonium Nitrate	Ammonium Sulfate	LSD
Constituents (%)	2021	CP	9.69	9.43	9.77	NS
		ADF	33.92	34.25	33.90	NS
		NDF	56.08	56.45	56.33	NS
	2022	CP	10.45	10.54	10.41	NS
		ADF	38.65	38.26	38.49	NS
		NDF	65.92	65.76	65.53	NS

Table 2. Crude protein (CP), Acid detergent fiber (ADF) and Neutral detergent fiber (NDF) values (%) for each rate of nitrogen application averaged across all sources in both trials. Least significant difference (LSD) is given unless the values were not significant (NS).

			Nitrogen Application Rate (lb/A)				
			0	30	60	90	LSD
Constituents (%)	2021	CP	9.42	9.11	9.59	10.40	NS
		ADF	33.03	34.20	34.60	34.27	0.908
		NDF	54.26	56.39	57.30	57.20	1.387
	2022	CP	10.25	10.37	10.56	10.68	NS
		ADF	38.17	38.20	38.40	39.10	NS
		NDF	65.32	65.45	65.70	66.47	NS

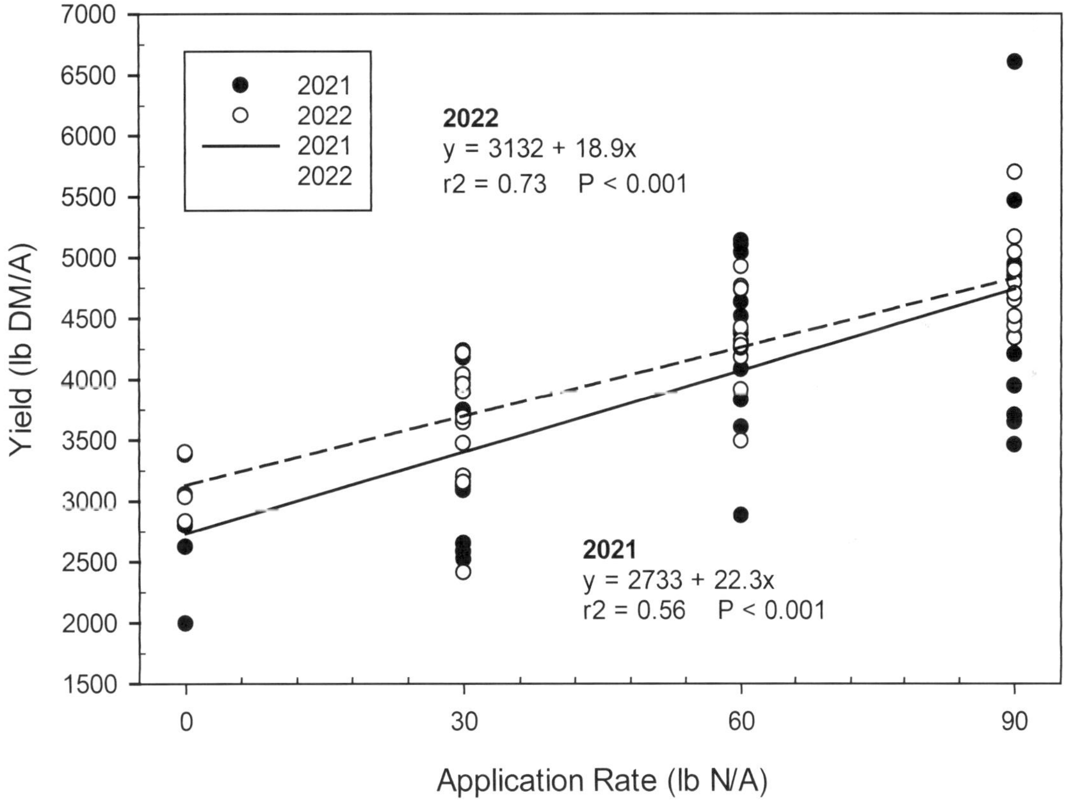

Figure 1. Yield (lb DM/A) data from all nitrogen sources graphed by application rate (lb N/A) for both trials. Polynomial regressions from each year are also shown together for comparison.

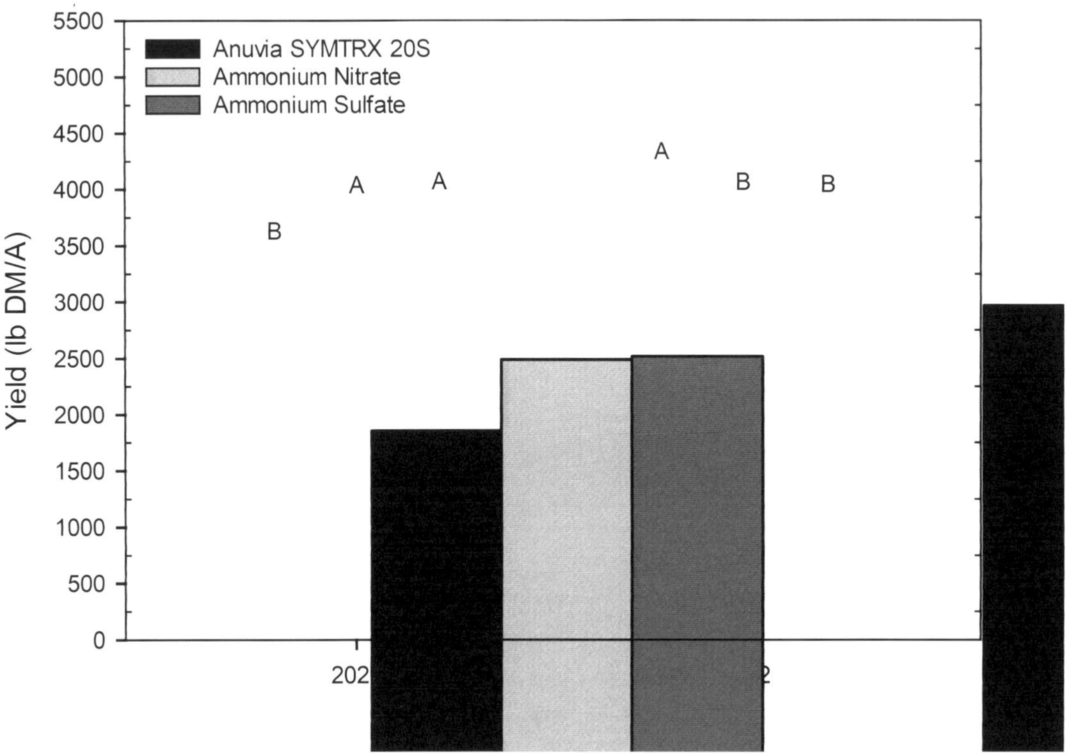

Figure 2. Comparison of average yield (lb DM/A) across all four application rates of each nitrogen source for both trial years. Bars within each year with the same letter designation are not different according to Fisher's protected least significant difference ($P < 0.05$).

IMPACT OF SEEDING RATE ON YIELD AND RETURN COST OF ANNUAL COOL-SEASON GRASSES

R. Lemus*[1], J.A. White[2], and J.B. Rushing[3]

Annual cool-season grasses (ACSG; annual ryegrass, small grains) have supplemented grazing over a wide range of soil and climatic conditions for the past 100 years. Our objective was to evaluate how seeding rate affects annual ryegrass and small grain forage mass, nutritive value, and vegetation dynamics in northeast Mississippi. Annual ryegrass ('Marshall) was planted at 20, 25, 30, 40, and 50 lb pure live seed (PLS)/acre. Small grains ('Wrens Abruzzi' cereal rye, 'Canmore' oat, 'Surge' triticale, 'Pembroke' wheat) were planted at 30, 50, 70, 90, and 110 lb PLS/acre. Urea (46-0-0) was split-applied when grasses reached three inches after establishment and after the first harvest. Treatments were harvested at 12 – 15 in height and subsamples were analyzed for nutritive value. Seasonal biomass of annual ryegrass was greater than small grains, and did not improve above 30 lb/acre seeding rate. Increasing the seeding rate of small grains improved forage biomass between 70 and 90 lb/acre. Seeding rate and species trends were also observed for Leaf Area Index and percent Green Canopy Cover. Nutritive value was affected by forage species; annual ryegrass had greater WS, but other comparable values to small grains. Preliminary results suggest that annual ryegrass seeding rates of 20 to 30 lb/acre and small grain seeding rates of 70 to 90 lb/acre optimize forage production, but could be influenced by planting date, establishment method, and fertilization. Seeding rates above that recommended for annual ryegrass and small grains did not reduce cost of production/ton DM.

[1]Professor and Extension Forage Specialist, Mississippi State University, MS (Rocky.Lemus@msstate.edu); [2]Forage Variety Testing Manager, Mississippi State University, MS (JWhite@pss.msstate.edu); [3]Associate Research/Extension Professor, Coastal Plain Experiment Station, Mississippi State University (BRushing@pss.msstate.edu).

INTEGRATED WEED MANAGEMENT REDUCES INPUT COSTS WHILE OPTIMIZING PRODUCTIVITY AND BIODIVERSITY IN MANAGED PASTURES.

Byron Sleugh, Scott Flynn, and Sam Ingram[1]

Feed costs represent approximately 50% of the cost of production in livestock systems. Improving productivity in terms of forage yield and quality could lead to lower production costs and lead to greater production of meat, milk, hay, or other revenue generating forage-based products. Weeds influence the structure and function of forage ecosystems, reduce feed value, decrease pasture carrying capacity, can be toxic or unpalatable to livestock, or cause other livestock health issues. Managing weeds requires diverse and integrated methods that favor forages or lead to a desired plant community with a resilient vegetation complex that can resist reinvasion. Therefore, it is important to consider implementing appropriate and integrated weed control strategies or methods when needed. Choosing the right weed control method or combination of methods (chemical, mechanical, cultural, biological) can reduce cost and frequency of weed control activities while improving forage productivity. Research has shown that when weeds are controlled as part of an integrated management approach (including soil fertility, grazing management, etc.), it is possible to: increase forage and livestock productivity, lower overall input costs and reduce the impact on animal health or animal performance.

[1]Global Leader, Weed Management Biology and Field Sciences, Corteva Agriscience, Indianapolis, IN (byron.sleugh@corteva.com); Zonal Biology Leader, Corteva Agriscience, Lee Summit, MO (scott.flynn@corteva.com); Field Scientist, Corteva Agriscience, Savannah, GA (sam.ingram@corteva.com).

COMPARISON OF WEB SOIL SURVEY AND KY GRAZE FOR ESTIMATING CARRYING CAPACITY

R.M. Kaplan-Fardy, J.C. Henning and S.R. Smith[1]

Carrying capacity is the number of livestock that can be kept on a farm without negatively affecting forage productivity and environmental sustainability. Web Soil Survey (WSS) and the KY Graze Model (KY Graze) are tools that can be used to estimate carrying capacity. The objective of this project was to estimate carrying capacity using both methods on five KY livestock farms ranging from 31 to 136 acres. WSS is an online tool developed by NRCS and uses estimated forage production based on USGS soil types to give an Animal Unit Month (AUM) rating. KY Graze is a spreadsheet-based model created by agronomists and grazing specialists from the Univ. of KY and the KY USDA-NRCS which estimates carrying capacity using forage species, management level and grazing efficiency in addition to soil types on each farm. For these comparisons, a 1,200 lb animal unit (AU) equivalent was used to estimate AU per farm. In all cases, WSS gave a higher carrying capacity compared to KY Graze. For example, on a 136-acre KY farm, WSS estimated that 78 AU could be maintained year-round on the farm in comparison to 57 AU with the KY Graze model. On a 124-acre KY farm, WSS estimated carrying capacity at 72 AU compared to the KY Graze at 38 AU. On both farms, KY Graze assumed a high management level and 80% grazing efficiency, but management and grazing efficiency are not adjustable with WSS. In summary, these results suggest that KY Graze is a more conservative method to estimate livestock carrying capacity and WSS may overestimate carrying capacity leading to unintended consequences.

[1]Reilly Morgan Kaplan-Fardy, MSc Student, (reilly.fardy@uky.edu), Jimmy Henning, Extension Prof., (jhenning@uky.edu), S. Ray Smith, Extension Prof., (raysmith1@uky.edu); Univ. of KY.

KNOTROOT FOXTAIL CONTROL IN BERMUDAGRASS HAYFIELDS

J.L. Belcher[1]

Knotroot foxtail (*Setaria parviflora*) is a warm-season perennial species that can occur from the Gulf Coast up into the Great Lakes region. Unlike annual foxtails, which reproduce only by seed, knotroot foxtail reproduces by both seeds and short, knotty rhizomes. These rhizomes make this species much more difficult to control. For the hay producer, foxtail seedheads lower hay quality and if present in hay, can cause ulcers in the mouths of horses and cattle. This study was conducted in Shorter, AL and initiated October 14, 2021 in order to evaluate applications of postemergence herbicides for control of knotroot foxtail and bermudagrass injury. Application was timed to be within one month of expected first frost. Treatments evaluated were 8, 12, and 16 fl oz/a of a 4 lb/gal glyphosate applied alone or with Pastora at 1.25 oz/A, Velpar L at 2 pints/A, Plateau at 6 oz/A, and Facet L at 32 oz/A. Rezilon, a preemergence herbicide, was applied at the time of initial application and again on February 16, 2022, in order to prevent any foxtail seed from germinating. This allowed for evaluation of control of the perennial plants without the influence of new seedling emergence. Treatments were applied to foxtail that had been cut and regrown to a height of 4 inches at the time of application as well as to a site that had not been cut and foxtail had grown to a height of 24 inches. Results show that applications of 12 fl oz/A or higher of glyphosate applied alone, or all rates in combination with Pastora provided good control of knotroot foxtail, regardless of height, at time of application. Velpar L, Plateau, and Facet L were ineffective at controlling knotroot foxtail when applied in the fall of the year.

[1]Regional Stewardship and Dev. Manager, Envu, Auburn, AL (jason.belcher@envu.com).

LEGUME ESTABLISHMENT IN NATIVE WARM-SEASON GRASS PASTURES

C.B. Phillips[1] and K.M. Payne[2]

Native warm-season grasses (NWSG) are well-adapted to the eastern United States and highly productive in the summer months, but forage quality and early season availability have limited their adoption by producers. These drawbacks may be improved by interseeding legumes into NWSG. Our objective was to evaluate the timing of seeding, the method of seeding, and the efficacy of burning residue on legume establishment in NWSG pastures consisting of a mix of big bluestem, little bluestem, and Indiangrass. Experimental plots were seeded in fall without burning, in winter without burning, or in winter with residue burned; using drill seeding (red clover at 4 lbs/ac, white clover at 1 lb/ac), broadcast seeding (red clover at 6 lb/ac, white clover at 1.5 lb/ac), or not seeded for a total of nine treatments with four replications. Emergence counts were taken in each plot at the start of the season, and species composition was determined before each harvest. In the first year of the study, the drill method had the greatest clover emergence, followed by broadcast, and then control, in both burn (27, 17.3, and 3.3 clover plants/ft^2, respectively) and winter (15, 10.3, and 0.5 clover plants/ft^2) treatments; however, seeding method was not significant when planted in the fall. Clover content in burn and winter treatments was affected by seeding method, with the drill method having the greatest clover content (36.8% and 18.2%, respectively) followed by broadcast (26.0% and 3.3%, respectively) and control (9.7% and 0%, respectively). Harvest yield was not affected by seeding timing or method. Plot composition will be measured for another year as emergence is repeated in new plots.

[1]Carter B. Phillips, MS Candidate, School of Plant and Environmental Sciences, Virginia Tech;
[2]Dr. Kathryn M. Payne, Applied Forage Systems Specialist, Southern Piedmont Agricultural Research and Extension Center, Blackstone, VA, Virginia Tech.

MINERAL CONCENTRATIONS OF CRABGRASS AS AFFECTED BY RATE AND SOURCE OF NITROGEN APPLICATION

K. R. VanValin[1], C. D. Teutsch[1], B. L. Hendrix[1], and W. M. Tilson[2]

Crabgrass is a warms season annual grass, considered a weed due to prolific growth and spreading morphology. However, crabgrass has potential to be a high-quality forage for summer grazing. When evaluating the nutrient content of forages, mineral concentration should be considered in addition to energy and protein. It is also important to consider how agronomic practices may affect mineral concentrations. This study evaluates mineral concentrations of crabgrass treated with 0, 50, 100, 150, 200, 250, or 300 lbs N/Ac from ammonium nitrate or broiler litter. Nitrogen application increased Ca, K, and Mg, and decreased P concentrations. Zinc concentrations were greatest in the 0-nitrogen group but lowest in the 50 and 250 lbs N/Ac groups. Manganese concentrations were similar for 0, 50, 100 lbs N/Ac groups but increased from 150-250 lb N/Ac and were greatest in the 300 lb N/Ac group. Copper and B concentrations were similar across all N treatments. Concentrations of Ca and Mn were greater for ammonium nitrate vs. broiler litter. Concentrations of P, Zn, Cu, and B were lesser for ammonium nitrate vs. broiler litter. Concentrations of K, Mg, and Fe were unaffected by N source. This suggests that crabgrass supplies mineral concentrations that exceed requirements for grazing beef cattle for 7 of the 9 minerals evaluated. However, availability of the mineral within crabgrass should be considered when developing mineral supplements for grazing livestock.

[1]Assistant Extension Professor, University of Kentucky, Animal and Food Sciences, Princeton, KY (Katie.VanValin@uky.edu); [2]Associate Extension Professor, University of Kentucky, Plant and Soil Sciences, Princeton, KY;[3]Senior Research Technician, University of Kentucky, Plant and Soil Sciences, Princeton, KY; [4]Virginia Tech (Retired).

MITIGATION STRATEGIES FOR HEIFER DEVELOPMENT ON TALL FESCUE

M. L. Boone, R. C. Carrell, S. L. Dillard[1]

Tall fescue *(Schedonorus arundinaceus)* contains a symbiotic endophyte that reduces animal performance and conception rates in cows and heifers. The objective was to mitigate the toxic effects of 'KY 31' tall fescue by limit-grazing cool-season annuals as an alternative to feed supplementation. In January 2022, 48 heifers were randomly assigned to tall fescue, tall fescue with supplemental feed, tall fescue with limit grazing of cool-season annuals for 24 h per week, and cool-season annuals. Body condition scores (BCS), hair coat scores (HCS), and body weight (BW) were recorded every 30 d. Forage samples were taken for biomass, nutritive value, and ergovaline concentration. On April 1, 2022, heifers were artificially inseminated (AI), and conception rates were determined after 45 d. Data were analyzed using the Glimmix procedure of SAS 9.4 (SAS Inst, Cary, NC) with $\alpha = 0.05$. Conception rates averaged 75, 67, 42, and 50% for the continuous, limit graze, supplemental feed, and tall fescue treatments, respectively. Crude protein (CP) concentrations for the control and limit grazing treatments were greater than the continuous grazing and the supplemental feed treatments (13.5, 10.7, 10.4, and 8.9%, respectively; ($P < 0.0120$). Neutral detergent fiber (NDF) and acid detergent fiber (ADF) were different among the supplemental treatment when compared to other treatments (48.2% vs 64.7% NDF and 25.8 vs. 37.0% ADF; $P < 0.0361$). The results indicate that limit grazing cool-season annual forages may be an option to mitigate fescue toxicosis.

[1]Graduate Research Assistant, Auburn University, Auburn, AL (mlb0108@auburn.edu); Beef Technical Services Specialist, Cargill Animal Nutrition, Nacogdoches, TX (Russell_Carrell@cargill.com); Assistant Professor & Extension Specialist, Auburn University, Auburn, AL (dillasa@auburn.edu).

NATIVE WARM SEASON GRASS GRAZING DEMONSTRATION

K.M. Payne, G.J. Pent, J.B. Daniel, and C.B. Phillips[1]

While tall fescue (*Schedonorous arundinaceus*) is the predominant forage species in Virginia pasture systems, the species has limited productivity during the summer months. Native warm-season grasses (NWSG) are well adapted to this region's climate and soils, maintaining high productivity with minimal inputs and may have the potential to support stocker cattle during this time. Between May and September of 2021 and 2022, 32 stocker cattle rotationally grazed either novel tall fescue or a mixture of big bluestem (*Andropogon gerardii* V.), little bluestem [*Schizachyrium scoparium* (Michx.)], and indiangrass [*Sorghastrum nutans* (L.) Nash] at the Virginia Tech Southern Piedmont Agricultural Research and Extension Center in Blackstone, Virginia where animal performance, forage production, and forage nutritive value were determined. Three full rotations of 28-35 days were achieved on these paddocks during each year. Although the average forage yield of the native warm season grasses throughout each growing season was greater than tall fescue (2,322 vs. 1,123 lb DM/ acre, respectively), livestock weight gains were more consistent among the two forage species. This may be explained by the greater forage nutritive value of the tall fescue forage; in 2021, unfertilized tall fescue forage averaged 13.9% crude protein and 57.0% neutral detergent fiber compared to 11.8% crude protein and 63.9% neutral detergent fiber of the native warm season grass mixture.

[1]Applied Forage Systems Specialist, Virginia Tech Southern Piedmont Agricultural Research and Extension Center, Blackstone, VA (kathrynpayne@vt.edu); Superintendent, Virginia Tech Shenandoah Valley Agricultural Research and Extension Center, Raphine, VA (gpent@vt.edu); Forage and Grassland Agronomist, USDA-NRCS, Farmville, VA (j.b.daniel@usda.gov); MS graduate student, Virginia Tech, Blackstone, VA (carterp00@vt.edu).

NATIVE WARM-SEASON FORAGE YIELD AND NUTRITIVE VALUES AS AFFECTED BY NITROGEN FERTILIZATION

Eric Bisangwa[1], Johnathan Richwine[2], Patrick Keyser[1], & Forbes Walker[1]

Although native warm-season grasses (NWSG) have been considered to have low fertility requirements, few data exist on their response to N, especially within the southeastern U.S. Experiments were conducted at East Tennessee (ETREC, Knoxville) and Highland Rim (HRREC, Springfield) AgResearch and Education Centers, 2015-2019. Three NWSG's (big bluestem [BB; *Andropogon gerardii* Vitman], switchgrass [SG; *Panicum virgatum* L.], and eastern gamagrass [EG; *Tripsacum dactyloides* (L.) L.]) were evaluated at each location to assess dry matter (DM) yield, and forage nutrient values responses to N (urea). Plots were harvested twice per season – once in June and a second time in either August or post-dormancy. Nitrogen was applied annually in a split application in late April and again following the first harvest at 0, 30, 60, 120, 180, and 240 lb N/acre N for BB and SG and 0, 60, 120, 180, 240, and 360 lb N/acre for EG. At both sites, N increased DM yield for all three species except at HHREC, where BB did not respond to N. Greatest producing N-rate ranged from 60 to 180 lb N/acre depending on species and site. All species responded to N with maximum yields at 120 lb/acre (BB and SG) or 180 lb/acre (EG). Responses on poorly drained soils (HRREC) were muted. Protein increased from 7.2, 7.7 and 7.3%, to 10.4, 10.8 and 12.6% for BB, SG, and EG, respectively, between 0 and 404 lb N/acre and TDN from 57.3, 58.0 and 56.9% to 58.9, 60.1 and 59.0% for BB, SG and EG, respectively. Native grass species being managed for forage production could benefit from 60-120 lbs ac^{-1} N, depending on N cost and production objectives.

[1]Center for Native Grasslands Management, School of Forestry, Wildlife, and Fisheries, Univ. of Tennessee, Knoxville, TN, [2]Arkansas State University.

YIELD AND NUTRITIVE VALUE OF SORGHUM-SUDANGRASS, COWPEA, AND SORGHUM-SUDANGRASS-COWPEA MIXTURES AMENDED WITH BROADCAST OR INJECTED POULTRY LITTER

C.C. Nieman[1], M.P. Popp[2], I. Thomas[3], K.P Coffey[3]

Poultry litter is a common fertilizer utilized by livestock farmers for pastures and hay fields in the mid-south. Subsurface application of poultry litter has improved nutrient utilization compared with broadcasting, potentially improving yield or nutritive value. This study compared yield and nutritive value of sorghum-sudangrass (SS), sorghum-sudangrass mixed with cowpea (SS-C), and cowpea (CW), when poultry litter was injected subsurface, broadcast, or not applied (control). Poultry litter treatments were applied on 6 June 2021 and forage species were drilled (off-set from subsurface bands) into killed bermudagrass sod on 6 June 2021 and on 23 May 2022. Plots were harvested on 8 Aug and 21 Sept in 2021, and 22 July and 22 Sept in 2022. Across years there was no interaction between forage species and poultry litter amendment type. Forage mass did not differ between injected (3.1 tons/acre) and broadcast treatments (2.5 tons/acre), though both were greater than the control (1.8 tons/acre). Forage yield did not differ between SS (3.7 tons/acre) and SS-C mixtures (3.2 tons/acre); CW was lesser and averaged 0.5 tons/acre. Crude protein (CP) was greater in CW (12.7%) compared to SS (9.7%) and SS-C (9.5%). Injected treatments (11.7%) had greater CP compared to broadcast (10.2%) and control (10.0%) treatments, which did not differ. No benefit to yield was observed for injected poultry litter treatments compared to broadcast, though, forages with injected poultry litter had increased CP concentrations compared to broadcast and control treatments.

[1]USDA-ARS, Dale Bumpers Small Farms Research Center, Booneville, AR; [2]University of Arkansas, Department of Agricultural Economics and Agribusiness, Fayetteville, AR; [3]University of Arkansas, Department of Animal Science, Fayetteville, AR

NORTH CAROLINA WILDLIFE RESOURCES COMMISSION'S NATIVE WARM SEASON GRASS PROGRAM

Jason Smith[1]

In 2001, North Carolina Wildlife Resources Commission (NCWRC) began the Cooperative Upland-habitat Restoration and Enhancement program (CURE) which established multiple focal areas across the state. Habitat enhancement of private lands included thinning and burning timber stands, establishing crop field borders, and converting fescue to Native Warm Season Grasses (NWSG) in pastures and hay fields. Establishment of NWSG for the dual purpose of livestock forage production and wildlife habitat was the most successful in terms of adoption and provided early successional habitat as well as quality forage for livestock during the summer forage gap. Although NCWRC no longer provides cost share to landowners, USDA Environmental Quality Incentives Program staff continue to provide technical guidance and specialty equipment to help livestock producers successfully establish and maintain NWSG. In 2019, current NWSG cooperators were surveyed; 92% felt these grasses were a very important part of their farming operation, and 85% of the livestock producers indicated that they were interested in incorporating more NWGS into their forage program. These results led NCWRC to increase outreach of the NWSG program, which included on-farm research concerning cattle weight gains and grazing days, and small ruminant performance. Average daily gain data indicated that NWSG provided a 1 lb/day advantage compared to typical fescue pastures. The NCWRC NWGS Program, beginning with CURE through the current program, has resulted in many landowner success stories and continues to gain popularity as livestock producers learn and see economic benefits that NWSG provide to their livestock and forage program.

[1]District 7 Wildlife Biologist, North Carolina Wildlife Resources Commission.

PRELIMINARY RESULTS: IMPROVING EASTERN GRASSLANDS WITH NATIVE WARM-SEASON GRASSES

M. Berry, P. Keyser, S. Jagadamma, H. Nauman, K. Coffey, C. Nieman, P. Owens, A. Ashworth, and E. Bisangwa[1]

Native warm-season grasses (NWSG) can produce high quality forage and high rates of gain for beef cattle. However, little data is available on how NWSG affect the productivity of cow-calf operations on a farm scale. Therefore, we implemented an experiment at three sites, Booneville, AR, Linneus, MO and Louisville, TN, with cow-calf pairs (mature cows over ≥3 years old, spring calving). We evaluated two forage systems that mix either a drought or drought/flood tolerant native C_4 species [big bluestem (BB) blend or eastern gamagrass (EG)] with a cool-season perennial, tall fescue (TF), and compared them to the most frequently used forage system within the Fescue Belt region, one that relies on TF only. The TN study site contains EG, with big bluestem at the MO site, and both big bluestem and EG at the AR site. Cattle (n = 12 pairs per experimental unit) were weighed yearly before initial grazing and again after final removal. Forage samples (n = 15) were collected at the beginning of grazing and once every twenty-eight days during the grazing season, and finally, at the conclusion of grazing. Harvested forages were tested for forage nutritive content (CP, NDF, ADF) using NIRS. Hay produced per forage system was documented by counting bales and weights of subsamples. The AR site was not able to participate in the first year of the study. Overall, there were no statistical differences between treatments in the first grazing season for either cattle or forage measures. However, cattle spent less time on NWSG in 2021 at the TN site to enable renovation of EG to be completed. Also, TF stands had a significant proportion of volunteer warm-season grasses within the pastures. Data from the second year of the study are currently under analysis.

COMPARING CALCIUM SOURCES TO NEUTRALIZE SOIL ACIDITY – A LABORATORY INCUBATION

E.L. Ritchey, C.D. Teutsch, and J.H. Grove[1]

Abstract

The foundation of a good soil fertility program begins with proper pH management. Acid soils are neutralized by the addition of limestone, primarily calcitic limestone. However, there is a common perception that calcium is responsible for the neutralization rather than the carbonates associated with calcium in the limestone. Some products containing calcium chloride are claiming a cheaper, faster, and more effective adjustment of soil pH compared to that of limestone. A laboratory incubation trial was initiated to research the effectiveness of three products in increasing acid soil pH value due to numerous requests from agricultural agents, producers and agribusiness representatives in Kentucky. The laboratory incubation trial utilized a Crider silt loam soil with an initial soil pH of 5.2. Treatments included an untreated check, liquid calcium (5 gallon acre^{-1}), pelletized lime (RNV 83), and ag lime (RNV 79). Both lime treatments were applied at 2 ton acre^{-1} after adjusting for RNV. Lime treatments were added to 1.7 oz (50 g) of soil and maintained moisture at 80% pore filled volume for the first 4 months. The cups were subjected to ambient conditions for remainder of the experiment. Soil pH increased for the pelletized lime and ag lime at each sample date relative to the untreated check and the liquid lime product. The untreated check and liquid calcium products did not change soil pH. This was not unexpected due to the inability of liquid calcium ($CaCl_2$) to consume acidity. To effectively neutralize soil acidity and increase soil pH, the addition of products that contain carbonates, oxides, or hydroxides must be utilized. Products that contain calcium or calcium chloride don't neutralize soil acidity. These findings agree with fundamental chemistry principles.

Introduction: Proper soil pH management provides the foundation to a good forage soil fertility program. Soil pH indicates the amount of active acidity present in the soil and influences nutrient availability, plant root growth and function, the rate of many biological processes and herbicide activity (Miller and Kissel, 2010). Lime application rate is based on the amount of active acidity and the soil buffer pH (Sikora, 2006). Application of acid neutralizing products are added to the soil to neutralize acid soil pH. The primary products used for pH management in field agricultural settings are some form of limestone, either calcitic or dolomitic carbonates. Some believe or claim that calcium is responsible for the neutralization rather than the carbonates

[1] Associate Extension Professor, University of Kentucky Research and Education Center, Princeton, KY (edwin.ritchey@uky.edu); Associate Extension Professor, University of Kentucky Research and Education Center, Princeton, KY (chris.teutsch@uky.edu); Professor of Agronomy/Soils Research & Extension, University of Kentucky Research and Education Center, Princeton, KY (jgrove@uky.edu).

associated with calcium in the limestone, but this is not true. Limestone application rates are adjusted according to their relative neutralizing value (RNV). The RNV is influenced by the purity and the fineness of the limestone. Some companies offer products that claim to neutralize acidity by adjusting the amount of base cations (Ca, Mg, K, or Na) present on the exchange complex by adding minute amounts of Ca in the form of calcium chloride ($CaCl_2$). To neutralize acidity, the proton (H^+) must be consumed. The neutralization reaction of calcitic limestone is demonstrated in equation 1.

Equation 1. $CaCO3 + 2H^+ \rightarrow H_2CO_3 \rightarrow H_2O + CO_2 + Ca^{2+}$

The acidity or proton in equation remains after the addition of $CaCl_2$ and has no liming ability according to equation 2.

Equation 2. $CaCl_2 + 2H^+ \rightarrow Ca^{2+} + 2H^+ + 2Cl^-$

Based on numerous questions and concerns from Kentucky producers, agribusiness, and agricultural producers we designed a laboratory experiment to test the effectiveness of liquid calcium chloride compared to traditional liming materials utilized in forage production.

Materials and Methods: A laboratory lime incubation study was established in the summer of 2021 at the University of Kentucky Research and Education Center (UKREC) in Princeton, KY and ended approximately one year later. The treatments included: an untreated check (nothing applied), liquid calcium at 5 gallon $acre^{-1}$, pelletized lime (RNV = 83), and agricultural lime (RNV 79). Both pelletized lime and ag lime were applied at 2 ton $acre^{-1}$ 100% effective lime, after adjusting the rate for the RNV of the product. The incubation study utilized soil from the surface 6 inches of a Crider silt loam (Typic Paleudalf) soil with an initial soil water pH of 5.2. Soil (50 g or 1.7 oz) was placed in 4 oz specimen cups with small holes in the cap to allow for gas exchange and replicated 4 times. Cups were maintained at 80% water filled pore space by weight until before the 6-month sample date. Every two weeks the cups were weighed, water was added by mass, and returned to the storage area. Two months before the 6-month samples were to be analyzed the storage area was destroyed by an EF-4 tornado. Many samples were recovered but cup moisture wasn't maintained after the tornado event and the 6-month and 12-month samples remained at ambient conditions. Destructive sampling occurred at 1, 3, 6 and 12 months. Results for the incubation study were reported as actual pH. Data was analyzed using SAS version 9.4 (Cary, NC).

Results and Discussion: Soil pH values responded to the applied treatments as expected according to equation 1 and equation 2. The initial soil pH value was 5.2 at the onset of the experiment prior to treatment application. One month after treatment application the pelletized lime and ag lime had significantly increased soil pH by 0.7 and 0.9 units from the initial soil pH (Table 1). Soil pH increased 1.1 and 1.0 units at the 3-month sample date for the pelletized and ag lime treatments. However, during the same time period the untreated check and the liquid calcium treatments were 0.1 to 0.2 units less than the initial soil pH (i.e. soil pH decreased with time). The short-term results of the incubation were very effective for the limestone products. Ideal soil and environmental conditions led to a rapid neutralization reaction of soil acidity in this incubation time.

Results up to the 6-month sampling date were very effective and illustrate how ideal, controlled laboratory conditions can actually improve the speed of a neutralization compared to those that occur in the field in ambient soil conditions. Conditions were maintained in the laboratory where the neutralization reaction was expected to proceed as fast as possible. For example, soil moisture was maintained at 80 pore filled volume, temperature was near room temperature, and gas exchange between the cups and atmosphere were allowed to occur. A one unit change in soil pH would not be expected in field settings due to less than ideal environmental conditions constantly present.

An EF-4 tornado destroyed the storage room where the samples were stored for this experiment on Dec 10, 2021. The samples that were recovered were collected and moved to another location but soil moisture and temperature was not maintained moving forward. The specimen cups dried and were exposed to greater fluctuations in ambient temperatures. This may have influenced the results of the 6-month and particularly the 12-month experimental results. The soil pH was still increasing at the 6-month sample time and resulted in an increase of 1.3 and 1.2 units with the pelletized and ag lime, respectively (Table 1). This was a slight increase in soil pH over the 3-month sample date, but not as much as expected based on the previous results. The soil pH for the untreated check and liquid calcium were 5.0, which is slightly lower than the original pH values at the initiation of the incubation and statistically lower than the soil pH values resulting from the calcium carbonate products. This is a clear indication that the liquid calcium product ($CaCl_2$) has no liming ability.

Soil pH values resulting from the 12-month sample date had increased over the initial samples but were not as high as the 6-month sample date, they had decreased slightly. The soil pH for the pelletized lime was 6.3 and was 6.2 for the ag lime at the 12-month sample date (Table 1). The untreated check and liquid lime had also remained between 5.0 - 5.1. These results could potentially be due to the soil drying and fluctuating soil temperature after the tornado event. These results would be more typical of what would be expected in field settings.

The addition of limestone to agricultural fields is not considered an immediate reaction, like an application of urea fertilizer, and a noticeable pH change isn't expected to occur immediately. Rather the change in soil pH in the field is expected to occur over several months or up to a few years (Ritchey and McGrath, 2022). The change in soil pH resulting from the laboratory incubation occurred more rapidly than would be expected in a field setting. Almost ideal environmental conditions were present for the first 4 months of the experiment, until the tornado event. After this time the reaction occurred more like in a field setting. This was particularly true for the 12-month sampling event.

Table 1. Soil pH values for the laboratory incubation trial for soil pH 1-month, 3-month, 6-month, and 12-month after treatment application. Initial soil pH was 5.2 for all treatments prior to treatment application.

	------------------------ Soil Laboratory Incubation Time ------------------------			
Treatment	1 month	3 month	6 month	12 month
Pr>F	(<0.001)	(<0.001)	(<0.001)	(<0.001)
Check	5.2 a[1]	5.1 b	5.0 a	5.1 a
Liquid Calcium	5.3 a	5.0 a	5.0 a	5.0 a
Pelletized Lime	5.9 b	6.3 d	6.5 b	6.3 b
Ag Lime	6.1 b	6.2 c	6.4 b	6.2 b

[1] Letters in the same column that are different indicate significant treatment differences at the 0.1 level of significance.

<u>Conclusions:</u>

Results of the laboratory incubation trial indicate that proven products and practices to neutralize soil acidity still hold true and are in agreement with sound chemistry foundations. The products that were expected to neutralize soil pH (i.e. those containing carbonates) did neutralize soil pH and increased soil pH within a given incubation period. Products not containing carbonates – liquid calcium ($CaCl_2$) have no mechanism to change soil pH and did not change soil pH in this experiment. In short, to effectively neutralize soil acidity and increase soil pH the addition of products that contain carbonates, oxides, or hydroxides must be utilized – not products that just contain calcium or chloride.

References

Miller, R.O. and D.E. Kissel. 2010. Comparison of soil pH methods on soils of North America. Soil Sci Soc Am J. 74:310-310

Ritchey, E.L. and J.M. McGrath. 2022. AGR 1 Lime and Fertilizer Recommendations: 2022-2023. Univ KY Coop. Ext. Service., Lexington, KY

Sikora, F.J. 2006. A buffer that mimics the SMP buffer for determining lime requirement on soil. Soil Sci. Soc. Am. J. 70:474-486.

THE EFFICACY OF DIFFERENT LIMING MATERIALS FOR KENTUCKY FORAGE PRODUCTION

E.L. Ritchey, C.D. Teutsch, and J.H. Grove[1]

Proper pH management is the foundation to a good soil fertility program. Soil pH influences nutrient availability, root growth and function, and nutrient availability. Just because an agricultural product contains calcium doesn't mean that it will change soil pH. A field study was conducted to test the effectiveness of three calcium products in raising soil pH were compared to an untreated check in acid soils. A field incubation was conducted at 16 field locations across Kentucky in cooperation with county extension agents. The application rates were 5 gallon/acre for the liquid calcium and 2 ton/acre for the pelletized lime and ag lime. The field study resulted in higher soil at 3-month and 12-month sample dates with ag lime and pelletized lime than the check and liquid calcium. Furthermore, there was no increase in biomass yield or hay qualities with any of the calcium products used in this study. The soil pH was not improved with the addition of calcium chloride and the results of this study support the chemical foundations associated with soil acidity neutralization reactions - calcium chloride doesn't neutralize acidity and calcium carbonates do neutralize soil acidity.

[1]Associate Extension Professor, University of Kentucky Research and Education Center, Princeton, KY (edwin.ritchey@uky.edu); Associate Extension Professor, University of Kentucky Research and Education Center, Princeton, KY (chris.teutsch@uky.edu); Professor of Agronomy/Soils Research & Extension, University of Kentucky Research and Education Center, Princeton, KY (jgrove@uky.edu).

EFFECTS OF COVER CROPS ON SOIL QUALITY

Elisa Donley[1], Samuel Princ[1], Paul Cook[1], Dr. Iryna McDonald[2] and Dr. Brian Sancewich[2]

Cover crops offer several benefits for the agricultural sector by increasing the level of soil nitrogen while improving overall soil health. Several cover crop varieties were planted on the University Farm-Fort Hays State University. Undergraduate researchers conducted ground based observations including stand persistence and biomass sampling. Data was collected and analyzed for soil nutrients, pH, organic matter, soil moisture, and soil health index. Biomass data was evaluated for nutritive values. The objective of this study was to determine: (1) the effectiveness of cover crop usage on overall soil health and nutrient improvement and (2) evaluate a dual-purpose cover crop mixture as a tool that increases soil health, biomass production, and forage use. The results indicate that cover crop as a forage provide, on average, 0.88 dry weight tons/acre of biomass with high forage nutritive values: Crude Protein (CP) levels were 28.1 – 30.5%, Acid Detergent Fiber (ADF) 21.6 %, Neutral Detergent Fiber (NDF) 35% and Relative Feed Value Index 191.5. Solvita test (CO2 respiration) is an indicator of carbon-sequestration where the greater the release of carbon the greater the soil reserve. Based on results it was determined that CO2 respiration increased relative to the control plot (before-planting). Carbon respiration values were 45.50, 66.30, 66.30, and 29.50 for sample plot 4, 3, 2, and 1 (control), respectfully. The Haney soil test and Solvita test were used to identify improvement in soil health. Overall soil test results indicate soil health was improved were cover crops were planted, the soil health calculation values were 10.10, 9.59, 8.59 and 4.7 (control).

[1]Undergraduate Researchers, Department of Agriculture, Fort Hays State University: Elisa Donley (erdonley@mail.fhsu.edu), Samuel Princ (sdprinc@mail.fhsu.edu), Paul Cook (pfcook@mail.fhsu.edu); [2]Research Mentors: Dr. Iryna McDonald (i_mcdonald@fhsu.edu); Dr. Brian Sancewich (brsancewich@fhsu.edu).

STUDY ON YIELD POTENTIAL OF CO_3 NAPIER BAJRA HYBRID UNDER SUB-TROPICAL SITUATIONS IN INDIA

Om Singh, Ayushi Singh[1], Ananya Singh[2], Anita Singh[3]

Abstract

A field trial on CO_3 was conducted at ICAR-IVRI, Izatnagar during 2013-14 to 2017-18 to study the effect of harvesting season/time on green, dry matter yield and Morphological characters. Data shows that yield of CO_3 was increased from first year 2013-14 to third year 2015-16 and decreased in fourth year of plantation of crop. Re-growth of ratoon crop was good as it was in first year crop, it may be concluded that crop can be grown up to 5-6 years. The crop yield of first year and second year as par with fifth year. Maximum green fodder yield 307.7 q/ha in 2015-16. In 5th year (2017-18), ratooning yield was at par 303.3 q/ha with 2nd to 4th year (2014-15 to 2016-17). Good package of practices are recommended for ratooning to get good yield as well as economic production for long time 5-6 years of ratooning. The crop CO3 produced maximum green fodder yield (300.7 q/ha) in August-September.

Keywords: Napier hybrid, Nutrients, Crop, Animal nutrition, rainfall, dry matter.

Introduction: Napier grass (*Penninsetum purpureum* Schumach.) is a fast growing perennial grass native to Sub-Saharan Africa that is widely grown across the tropical and subtropical regions of the world. It is a multipurpose forage crop, primarily used to feed cattle in cut and carry feeding systems. Characterization and diversity studies on a small collection of Napier grasses have identified a moderate level of genetic variation and high-lighted the availability of some good agronomic expertise on precise phenotyping, genotyping and the application of molecular technologies to Napier grass improvement using modern genomic tools which have been applied in advancing the selection and breeding of important food crops.

Sustainable livestock production is highly dependent on the availability of quality feed and forage resources. Napier grass, also known as elephant is one of the most important tropical forage crop. It is widely used in cut and carry feeding systems. and is of growing importance in other agricultural systems. Napier grass possesses many desirable characteristics, including high yield per unit area, tolerance to intermittent drought and high water use efficiency, making it a forage of choice. It has the ability to withstand repeated cutting and will rapidly regenerate, producing palatable leafy shoots. Consequently, enhancing the knowledge-based use and conservation of the available Napier grass resources promises to substantially benefit livestock value chains.

Elephant grass, as implied by its name, is important source of forage for elephants in Africa, Elephant grass is analysis plant. Hybrid Napier CO_3 grass is one of the widely cultivated fodder grass because of its high nutritive value. This grass is an interspecific hybrid between

ICAR-IVRI, Izzatnagar, SVPUAT [1-2]; Modipuram, [3]KVNER, Bareilly (corresponding author: omsingh1964@gmail.com).

elephant grass and bajra grass. Compared to other fodder crops hybrid. This grass is called as "King of Napier grass". Super Napier grass is best suited to higher rainfall areas, but it is drought-tolerant and can also grow well in dry areas. It does not grow well in waterlogged areas. Hybrid Napier is a perennial grass which needs to be retained in the field for more years to reduce crop cultivation costs. Compared to Napier grass, Hybrid napier produces larger and softer leaves. Its ease of regeneration and high productivity make it perfect for the feeding of cattle, amongst other crop eating creatures.

Materials and Methods: A field trial was conducted on one acre of land of fodder form of ICAR-IVRI, Izatnagar, Bareilly (UP), India during 2013-14 – 2017-18. The crop CO_3 was planted in rows 60 x 60 and plant to plant 50 x 50 cm. Soil were sandy loam, medium in M, P and K availability and pH 7.2, water holding capacity, soil structure, soil texture and organic matter content (-1) needs to be analyzed. Harvesting of crop was scheduled at 60 days. As Ti- April-May (Spring), T2–June-July (Summer), T3- August-September (Rainy), T4- Oct-November, T5-December–January-February-March (Lean Period/Winter Season). Data were collected under RBD with three replications. Samples were analyzed for yield, dry matter and morphological characters. All recommended agronomic practices such as manure and fertilizer applications, weed management, plant protection including irrigation at moisture deficit. Organic manure 5t ha-1 was applied of the time of field propagation. Fertilizer doses maintained at every year.

Results and Discussion: The data shows that the morphological characters, green forage and dry matter, yield were significantly influenced with the year of cultivation and season/months of harvesting.

Effect on Morphological parameters

The plant height, leaf length, leaf width, number of leaves per culm and number of tillers per clump are contributing traits to forage biomass were affected significantly during the time of harvest. Harvesting during July-August-Sep these trait increased significantly as compare to harvesting months during December-January-February. Availability of rains/ moisture and long days hours, high temperature might influenced the plant physiology, higher photosynthates accumulation, better availability of moisture in root zone and reduction in abiotic plant stress conditions, along with sufficient availability of plant nutrients near root tips might be the reason to produce the significantly higher measures of the above traits.

Table 1 : Forage yield qha^{-1} of CO_3 as influenced by harvesting time and ratooning years

Treatment	2013-14	2014-15	2015-16	2016-17	2017-18
Apr-May	230.1	225.5	220.2	198.3	182.5
Jun-Jul	260.3	283.2	272.3	207.6	197.4
Aug-Sep	272.4	301.2	307.7	278.8	281.8
Oct-Nov	225.1	285.3	284.4	184.3	173.3
Dec-Feb-Mar	111.2	188.4	109.6	123.4	107.4
CD at 0.5%	7.32	12.34	9.24	7.12	6.18

Fig. 1 : Forage yield qha^{-1} of CO_3 as influenced by harvesting time and ratooning years

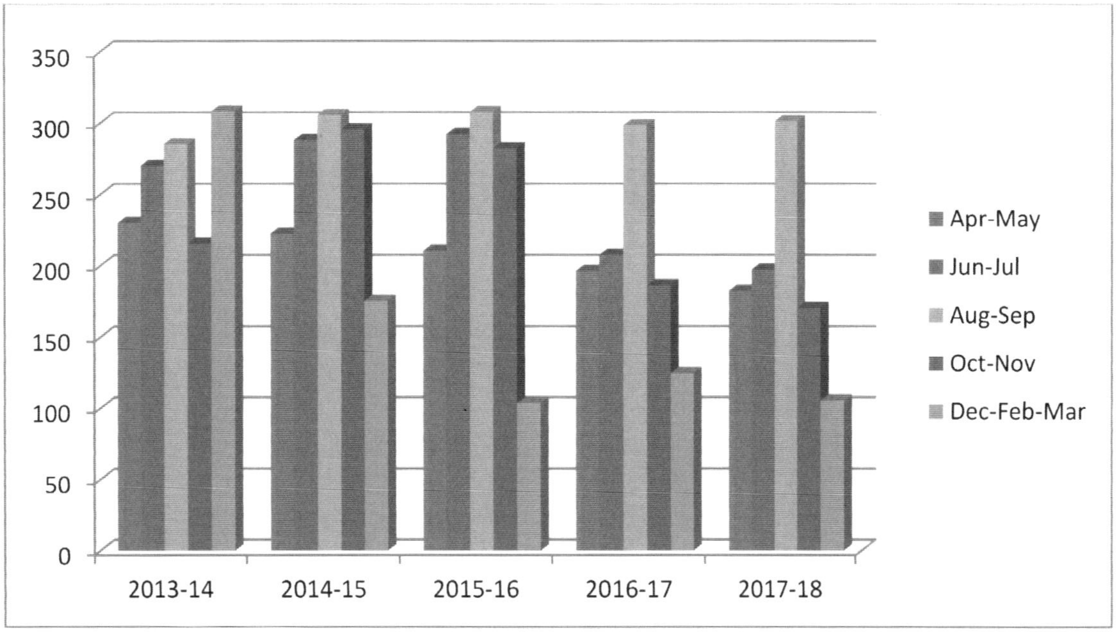

During the December-January-February due to low temperature (minimum and maximum) and short days hours in the Rohilkhand region increased plant stress on CO_3 crop and enhanced reduction in traits measurement.

Effect on Forage yield:

Forage yield was recorded higher while crop was harvested in summer and rainy season (230.1) as compared to winter months (108.2). This period may be called as lean period for forage production of CO3. This significantly reduction is yield might be due to favorable and unfavorable conditions to the plant. CO3 is tropical climate loving plant but in Rohilkhand Region falls under the sub-tropical climatic conditions and low temperature (minimum – below 10^0C and maximum below 20^0C) in December, January, February where is plant growth was found dormant. Hence, harvesting is suitable in last week and of March month. During this winter period of short duration inter crop like Barley, Berseem, Mustard or pea can be recommended to boost the ratoon crop from December to February. This practiced may increase CO3 forage yield and quality (leguminous- Barley/lucerne).

Table 2:Morphological characters of CO_3 at harvesting time/year at farm 2014-2018

Treatment	Plant height (cms)	No. of leaves per clump	No. of tillers per clump	Stem leaf ratio	Leaf length (cm)	Leaf width (cm)
Apr-May	50.4	93.6	18.9	0.71	73.1	3.2
June-July	52.6	105.3	27.3	0.75	94.3	3.4
Aug-Sep	57.1	107.4	33.2	0.81	100.4	4.8
Oct-Nov	44.2	102.1	32.2	0.81	103.2	4.1
Dec-Mar	35.7	85.3	29.1	0.93	91.6	3.3
CD at 0.5%	0.32	2.24	3.26	0.11	3.13	0.78

Fig. 2: Morphological characters CO_3 at harvesting time/year at farm, 20014-18

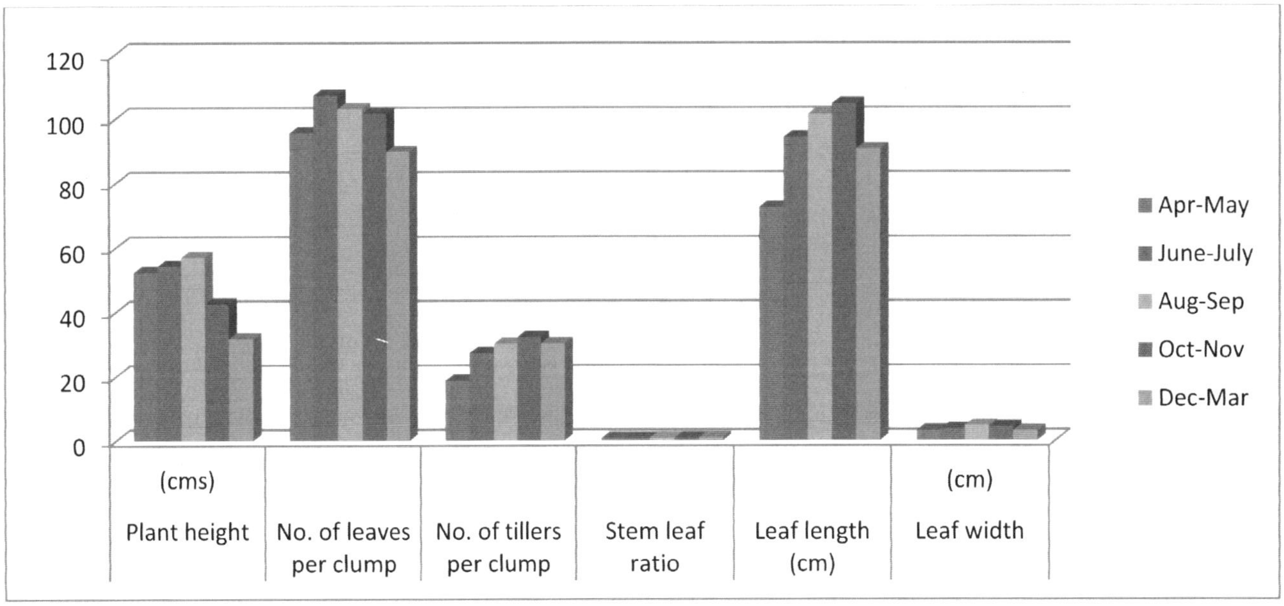

Effect of Nutrient management and Ratooning/regeneration:

Yield of crop was increased up to three years from 2013-14 to 2015-16 and found stable during two years from 2016-17 to 2017-18. The nutrition of crop is important besides NPK application of Bio-manures, FYM, vermi-compost, Bio-fertilizers, or organic manure. Mixing in inter rows is necessary to maintain regrow with and yield of 5 years ratoon crop. To control weeds in rows after rainy season sowing of inter crop like mustard, berseem, Barley is recommended. This crop may be incorporated in February-March by weeder or tractor tiller which will increase organic matter in the soil and availability of nutrients. Pea, Central, Menthi, or Palak will best the soil structure and fertility. Under trial Berseem was found better than

Annual Rye Grass and Barley when there were sown in November to February in rows to control weeds.

Table 3: Dry matter yield (DM) as influenced by harvesting time (years) :2014-2018

Treatment	2013-14	2014-15	2015-16	2016-17	2017-18
Apr-May	64.2	69.7	71.3	68.2	66.0
Jun-Jul	75.4	80.1	83.4	82.8	67.4
Aug-Sept	78.6	83.6	87.6	83.3	78.7
Oct-Nov	72.4	76.4	70.8	78.2	73.1
Dec- Mar	62.1	65.9	67.4	61.6	70.5
CD at 0.05%	3.25	3.16	3.31	5.09	4.9

Fig. 3: Dry matter yield (DM) as influenced by harvesting time (years): 2014-2018

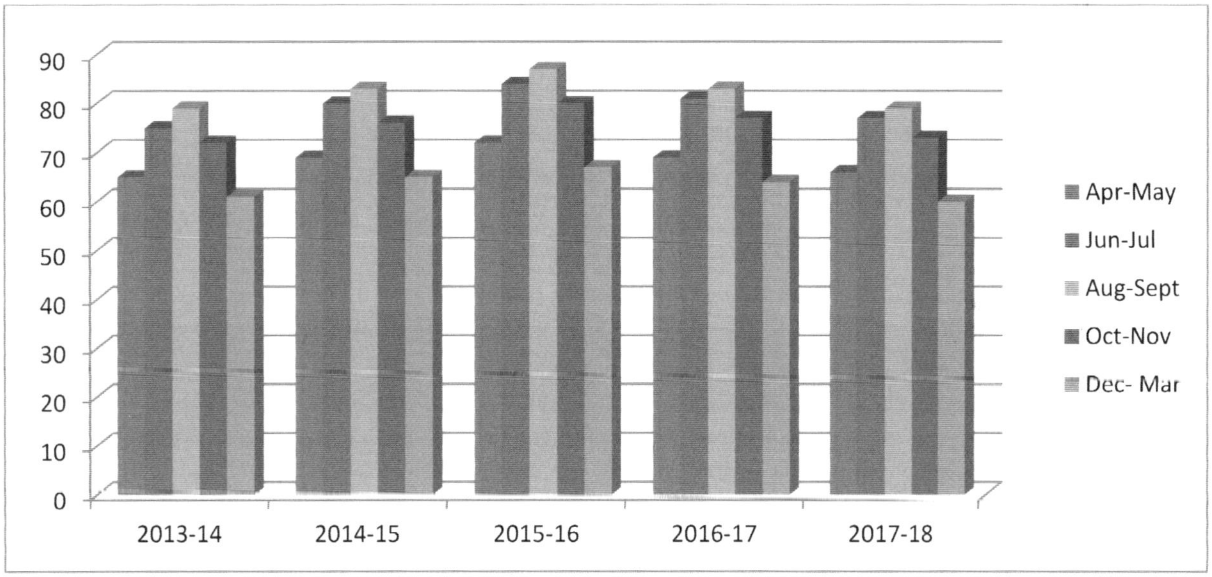

Effect of irrigation on Ratoon Crop of CO_3:

CO_3 was found better when soil moisture was maintained throughout the year. Sufficient water in rest zone provided abiotic stress free environment. To take ratoons of CO_3 for five years every month irrigation is important. It was found that where irrigation was delayed or Nutrition was not applied at time the growth and yield was reduced.

Effect of weed management:

During five years of crop ratooning it was found that culms of crop increased every year. Number of tillers per culm were maximum in 2017-18. To maintain yield potential, it is recommended that every year plants become thinner than first year. Therefore, proper weeding inter culture of crop, maintain and irrigation is necessary. Weeds compete with crop and reduce re-growth of tillers.

Effects of Harvesting on yield and forage quality:

Harvesting at right time is important. If harvesting is delayed every week crop become harder, unpalatable and increase in dry matter. Sweenlant leaves and stems are essential to increase digestibility. During rainy season crop growth faster than winter and summer season. Delayed crop found high in dry matter and indigestible content like lignin. During summer April-May-June crop shows ear head appearing. Reproductive stage/seed formation stage is a sign of physiological matter of crop. Therefore, roots with balanced nutrition, Irrigation weed free conditions are pre requisite of ratoon crop for five years of evaluation. Cutting with proper equipment manual on mechanical method is necessary along with proper height of plant at right time as well as proper water and fertilizer application after every harvest.

Conclusion: Data shows that yield and quality of forage of CO_3 was affected by several factors like harvesting time, harvesting year, integrated nutrient and weed management, along with maintenance of regime of water in root zone according to season especially during February-March-April-May and June before rainy season and post rainy season like October-November too. Plant is fast growing and needs frequent nutrition & water applications without proper percentage of agronomic practices crop may not sustain for five years. Because uptake of nutri boots and water is higher due to high biomass (1500 q /ha/year) production over the other multicut crops. Hence, crop with good package of practices can be grown upto 5th years as a ratoon of Co3 Napier Hybrid.

References:

1. Das, SK, Sharma, KL and Singh, Om (1995). Phosphorus and Sulphur availability in soil following in conjunction with various organic residues. Journal of the Indian Society of Soil Sciences (2), 223-228.
2. Om Singh, AKS Tomar and Hriom Pandey (2020). Napier Hybrid Evam Bajra ki Kheti. Extension Bulletin, Poster, Research Gate profile, SCSP, ICAR, April, 9.
3. Om Singh(2022), Sankar Napier Bajra Utpadan, Technical Bulletin, Poster, Research Gate profile.
4. Singh, Om, Chaturvedi, VB, Verma MR and Singh HC (2018). Effect of Integrated Nutrient Management on Hybrid Napier Production under irrigated conditions, International Journal of current Microbiology and Applied Sciences 7 (04) : 2731-2737.
5. Singh, Om; Verma, M.R. (2018). Effect of row arrangement on Sorghum-Cowpea intercrops in irrigated conditions, Archives, 18(1), 7-10
6. Singh, Om (2014). Effect of Integrated Nutrient Management on Rye Grass Production under irrigated conditions, Journal of Tropical Agriculture, 32(3-4), 857-860.
7. Singh, Om (2014). Effect of Inorganic source of Nutrients on Yield in Rice-wheat sorghum, Journal of Tropical Agriculture, 32(3-4), 851-855.

THE EFFECTS OF GRAZING COVER CROPS ON ANIMAL PERFORMANCE AND SOYBEAN PRODUCTION

J.B. Rushing[1] and J.C. Lyles[2]

Integrated crop-livestock systems (ICLS) are designed to produce both cash crops and livestock simultaneously on the same land base, which can reduce nutrient concentration and environmental risk, improve soil structure and function, exploit natural pest control processes, and increase overall farm productivity and economic sustainability. A field trial was conducted at the Coastal Plain Branch Experiment Station in Newton, MS on a Prentiss fine sandy-loam soil during the 2019-2021 growing seasons. The trial consisted of a randomized complete block design with three replications. Cover crop treatments included oats (O), O + crimson clover (OC), and OC + radish (OCR). Cover crop production (FM, CP, and TDN), soybean grain yield, animal performance (ADG, GAIN) and economic comparisons were made from each cover crop system. Cover crops differenced in nutritive analysis; O had significantly lower mean CP, but greater TDN. The OCR treatment had significantly lower ADG (3.03 lb hd^{-1} d^{-1}) from grazing beef steers than the other two cover crop treatments. Soybean grain yield was unaffected by cover crop species ($P > 0.05$). Net economic returns above crop production costs were lower for the OCR treatment (\$475.98 ac^{-1}) compared to the other two cover crop treatments. Findings suggest grazing oats or a combination of oats and crimson clover as a cover crop with subsequent no-till soybean production has the potential to increase net returns for livestock/row crop producers in east-central Mississippi.

[1]Associate Research/Extension Professor; Coastal Plain Branch Experiment Station, Newton, MS; [2]Research Associate I; Coastal Plain Branch Experiment Station, Newton, MS.

USING FARM DEMOS AND PASTURE WALKS TO PROMOTE NATIVE WARM-SEASON GRASSES FOR GRAZING

J.B. Daniel[1]

Abstract

The establishment and use of native warm-season grasses (NWSG's) for grazing livestock has resulted in limited adoption in the mid-Atlantic and the southeastern United States. Over the last 40 years there have been many research studies on the establishment and management of NWSG in grazing livestock systems however, until recently, there has been a relatively low level of interest by mainstream farm operations in Virginia.

In the early 1990's, Dr. Dale Wolf with Virginia Tech as well as Dr. Pat Keyser and Marc Puckett, at that time with the Game Department, encouraged interested farmers like Tim Willis, Greg Wade, Bob Kane, and several others to plant these NWSG's (primarily switchgrass) and manage them as a productive forage for grazing during the summer months. Despite the success of these farmers mentioned above, many other farmers were not interested due to the estimated long establishment period, or the risk of failure and the lack of confidence in these as forages by many of the trusted agricultural professionals in localities across Virginia. Other farmers would read popular press articles but there were very few opportunities to see and learn about these forages at demo sites, much less on farms in production systems.

In 2018, USDA began the Working Lands for Wildlife for Northern Bobwhite Quail program in Virginia, which provided a financial incentive for farmers to convert some pastureland to the wildlife-friendly NWSGs for grazing. To overcome the limited interest from the past, a proactive educational and awareness program was developed and promoted to educate the agricultural professionals and provide outreach to producers across Virginia. These included training sessions on field preparation and a 1-year establishment protocol for NWSGs to local conservation and Extension staff as well as pasture walks highlighting the successful establishment and productive use of these grasses on working farms in different regions of the state.

This education and outreach effort helped build some confidence in these NWSGs with local conservation and Extension agents while successfully generating more interest from producers across the state. While select farmers were signing up to plant NWSG's through the Working Lands for Wildlife Program in VA, it was obvious that more research and demonstration was needed across VA. USDA-NRCS in Virginia awarded a Conservation Innovation Grant (CIG) to Dr. Gabe Pent at Virginia Tech's Southern Piedmont Ag Research and Extension Center to establish a NWSG grazing demonstration site and cultivar trials of the four tall native grasses promoted for grazing at multiple locations across the state. During this

[1]USDA-NRCS State Grassland Agronomist and Grazing Specialist

time, NRCS also partnered with Virginia State University at Randolph Farm to renovate an old grazing demonstration area to establish paddocks with switchgrass, big bluestem and indiangrass, and eastern gamagrass.

These strategic partnerships resulted in successful establishment of NWSGs on grazing demonstration sites accessible to producers and highlighted on research farm tours annually in the Piedmont, Valley, and Southeast regions of Virginia. Several of the farmers who attended the educational events from 2018 have successfully established paddocks with these productive NWSGs as part of their grazing systems. These farmers are hosting pasture walks as outreach events in their local regions of the state to share farmer-to-farmer how these forages complement the cool-season pasture and benefit the overall production of their forage and livestock system.

Each year, farmers in each region of Virginia are converting more pasture acres to these productive grasses. With USDA's emphasis on Climate Smart Agriculture, using NWSG as part of a well-planned grazing system offers producers a practical solution to limiting their risk while building resiliency in forage systems across Virginia and much of the United States.

In Virginia, there is a growing network of producers who have successfully established NWSGs and are willing and available to host other interested farmers to see the grasses in person and ask their questions to a peer who has done it themselves. Despite advances in technology and the wonderful educational aspects of videos, webinars, and online meetings, nothing takes the place of taking an interested farmer to the field so they can see and touch these grasses, and watch cattle graze these grasses and ask their questions to an experienced farmer.

This type of coordinated approach to educating local agricultural specialists and providing outreach to farmers across the state is resulting in a renewed interest of establishing NWSG in forage systems across Virginia. The same system can be replicated in other states and regions. Begin by identifying strategic partners, looking for CIG's or other funding opportunities to establish grazing demonstrations, educating the local agricultural professionals, continually highlighting producer success stories, and connecting interested farmers with producers who have succeeded themselves. Finally, be ready to provide the detailed technical assistance to support these farmers with a conversion schedule including field preparation, planting dates, seeding rates, adapted cultivars, weed control and future grazing management. Part of their success will depend on your availability to answer their questions and provide the needed support with accurate and credible information along with encouragement throughout the process.

For opportunities to help replicate these efforts and expand the establishment of NWSG's to store more carbon in the soil and build resiliency in productions systems across the country, look for the $30 million USDA-funded project titled, Climate-Smart Grasslands – the Root of Agricultural Carbon Markets, led by Dr. Pat Keyser at the University of Tennessee, to roll out in 2023 across VA, NC, SC, AL, TN, KY, IN, MO, and AR. This project will focus on improved grassland management, including development of complementary forage systems that include both cool-season forages and NWSG. Additional funding opportunities are available in 24 states through the Working Lands for Wildlife Program landscape initiatives. There has never been a better time with as many incentives to establish native grasses on working lands!

USING TALL FESCUE SYMBIOTIC DIVERSITY TO IMPROVE PASTURE RESILIENCE AND SUSTAINABILITY

A.A. Jacobs[1], J.A. Nelson[1], A.E. Carlisle[1], M. Bagherzadeh[2], H. Ji[3], V. Stanton[4], and R.L. McCulley[1]

Tall fescue-dominated pastures cover 14 million ha in the eastern US and support 8.5 million beef cattle, making continued productivity and sustainability a high priority. Most tall fescue stands are infected with the symbiotic fungal endophyte *Epichloë coenophiala*, which improves grass growth and productivity in hot weather and under heavy grazing. Compounds produced by the most common tall fescue cultivar and strain of the fungus ('KY 31' common toxic strain) are harmful to grazing animals and lead to economic losses. Alternative fungal strains that do not produce alkaloid compounds toxic to grazing animals (novel strains) have been commercially available to producers for over a decade. Though individual tall fescue-novel endophyte symbioses have been evaluated for forage production and stand longevity, there is little information about the performance of mixed stands that contain multiple tall fescue cultivar endophyte symbiotic combinations. Using custom seed mixes, we established six levels of increasing symbiotic diversity and compared plant and soil parameters after five years. Symbiotically diverse stands produced more biomass in the fall and limited the invasion of other weedy grass species, while increasing soil carbon and nutrient pools. The most symbiotically diverse stand emitted more soil carbon dioxide than the least-diverse stand, perhaps due to greater root growth. These results show that increasing symbiotic diversity represents a new, readily adoptable, and effective strategy for increasing soil fertility and tall fescue productivity.

[1]Dept. of Plant and Soil Sciences, University of Kentucky, Lexington, KY; [2]National Wildlife Federation, Lexington, KY; [3]Kentucky Tobacco Research and Development Center, University of Kentucky, Lexington, KY; [4]Applied Statistics Laboratory, University of Kentucky, Lexington, KY. Corresponding author: alayna.jacobs@uky.edu

MINERAL CONTENT OF CRABGRASS AS AFFECTED BY NITROGEN APPLICATION RATE AND SOURCE

K. R. VanValin[1], C. D. Teutsch[1], B. L. Hendrix[1], and W. M. Tilson[2]

Abstract

Crabgrass is a warms season annual grass, considered a weed due to prolific growth and spreading morphology. However, crabgrass has potential to be a high-quality forage for summer grazing. When evaluating the nutrient content of forages, mineral concentration should be considered in addition to energy and protein. It is also important to consider how agronomic practices may affect mineral concentrations. This study evaluates mineral concentrations of crabgrass treated with 0, 50, 100, 150, 200, 250, or 300 lbs N/Ac from ammonium nitrate or broiler litter. Nitrogen application increased Ca, K, and Mg, and decreased P concentrations. Zinc concentrations were greatest in the 0-nitrogen group but lowest in the 50 and 250 lbs N/Ac groups. Manganese concentrations were similar for 0, 50, 100 lbs N/Ac groups but increased from 150-250 lb N/Ac and were greatest in the 300 lb N/Ac group. Copper and B concentrations were similar across all N treatments. Concentrations of Ca and Mn were greater for ammonium nitrate vs. broiler litter. Concentrations of P, Zn, Cu, and B were lesser for ammonium nitrate vs. broiler litter. Concentrations of K, Mg, and Fe were unaffected by N source. This suggests that crabgrass supplies mineral concentrations that exceed requirements for grazing beef cattle for 7 of the 9 minerals evaluated. However, availability of the mineral within crabgrass should be considered when developing mineral supplements for grazing livestock

Introduction: Crabgrass (*Digitaria* spp.) is a warm season annual grass, commonly considered a weed due to prolific growth and spreading morphology. However, crabgrass has significant potential to provide high-quality summer forage for grazing livestock in the transition zone. Improved varieties of crabgrass have shown to produce 20-25 lbs/acre (Dalrymple, 1999). Crabgrass treated with N from either ammonium nitrate or broiler litter exhibited crude protein concentration ranging from 7.6-10.5% (Teutsch et al., 2005). An additional benefit to crabgrass is its ability to re-seed itself, acting as a perennial (Dalrymple, 1999).

In addition to evaluating the energy and protein concentrations of a particular forage, it is also important to consider mineral concentrations. Minerals are a required nutrient for grazing livestock, and work to support major metabolic processes including growth, reproduction, lactation, and health (Suttle, 2010). Mineral requirements have been established for beef cattle and are designed to prevent deficiency. Additionally, some minerals have established maximum tolerable concentrations to prevent toxicities from occurring (NASEM, 2016). This study seeks to evaluate the mineral concentrations of crabgrass treated with varying concentrations of N supplied as either ammonium nitrate or broiler litter.

---

[1]Assistant Extension Professor, University of Kentucky, Animal and Food Sciences, Princeton, KY (Katie.VanValin@uky.edu); [2]Associate Extension Professor, University of Kentucky, Plant and Soil Sciences, Princeton, KY;[3]Senior Research Technician, University of Kentucky, Plant and Soil Sciences, Princeton, KY; [4]Virginia Tech (Retired).

Materials and methods: 'Red River' crabgrass (*Digitaria ciliaris* (Retz.) Koal) was established over two years (May 7th, 2001, and April 16th, 2002) near Blackstone, VA. The soil series for both years was a Wedowee sandy loam (fine kaolinitic, thermic Typic Kanhapludults). Initial soil nutrient levels in 2001 were 6.2 pH, 17 ppm P (high), 24 ppm K (medium -), 144 ppm Ca (Low +), and 40 ppm Mg (High -), and in 2002 were 6.2 pH, 35 ppm P (very high), 88 ppm K (high -), 287 ppm Ca (Low +), and 74 ppm Mg (high -). Plots were seeded by a cultipacker type seeder with a seeding rate of 6.0 lb/acre. Nitrogen was applied at 0, 50, 100, 150, 200, 250, and 300 lb plant available N per acre as a single application of either ammonium nitrate (AN) or broiler litter (BL). Nutrient content of broiler litter in 2001 was 51% dry-matter, 5.92% total kjeldahl nitrogen (TKN), 1.53% NH_4-N, 4.40 organic-N, 4.02% plant available nitrogen (PAN = NH_4-N × 0.90 + Organic N × 0.60), 2.41% P, and 3.48% K. In 2002 broiler litter was 49% dry-matter, 3.57% TKN, 0.77% NH_4-N, 2.80% organic-N, 2.37 PAN, 2.49% P, and 4.74% K.

Nitrogen treatments were immediately incorporated by disking once with a finishing disc (Virginia Department of Conservation and Recreation, 1995). Phosphorus and K were applied in both years such that all plots received the same amount contained in the greatest application of broiler litter. Bentazon and carfentrazone-ethyl herbicides were applied for the control of broadleaf weeds. Plots were 3 × 6.5 yards. Plots were harvested on July 5th and August 31st in 2001 and July 11th, August 15th, and October 3rd in 2002, when forage reached late boot stage, except for the last harvest of each year, which was harvested at seed maturity. A subsample of fresh forage was collected and used for mineral analysis.

Results and Discussion: Calcium concentrations ranged from 0.28-0.39% and increased with N concentration, and were greater for AN vs. BL ($P \leq 0.05$). Phosphorus concentrations ranged from 0.34-0.42% and decreased with N concentration, but were lesser for AN vs. BL ($P \leq 0.05$). Potassium concentrations increased with N application and ranged from 0.29-0.37% ($P \leq 0.05$). Potassium concentrations were similar between AN and BL ($P \geq 0.05$). Magnesium concentrations increased with N application ($P \leq 0.05$), and ranged between 0.36-0.53%, but were similar between AN and BL ($P \geq 0.05$). Zinc concentrations were greatest in the 0 Nitrogen group but were the lowest in the 50 and 250 lb/acre groups ($P \leq 0.05$). Zinc concentrations ranged from 24.7 to 27.0 ppm and were lesser in AN compared to BL ($P \leq 0.05$). Manganese concentrations were similar for the 0, 50, 100 lb/acre groups but were increased from 150-250 lbs/acre, and was the greatest in the 300 lb/acre group ($P \leq 0.05$). Manganese concentrations ranges from 78.7-136.7 ppm, and were greater in AN compared to BL ($P \leq 0.05$). Copper concentrations were similar across all N rates ($P \geq 0.05$), and ranged from 10.1-10.9 ppm. Copper concentrations were lesser in AN compared to BL ($P \leq 0.05$). Iron concentrations were the lowest in the 50 lb/acre group, were intermediate in the 0, 100, 150, 250, and 300 lb/acre groups, and were the greatest in the 200 lb/acre group ($P \leq 0.05$). Iron concentrations ranged from 56-65.8 ppm, and were similar between AN and BL ($P \geq 0.05$). Boron concentrations were similar regardless of N application rate ($P \geq 0.05$), but were lesser in AN compared to BL (*P* £ 0.05).

The nutrient requirements for beef cattle differ by stage of production (i.e. growing cattle, dry cows, lactating cows etc). Macro mineral requirements are impacted by stage of production; however, requirements for trace minerals have not been refined to a point where differences exist between stages of production. One exception to this would be Mn which is 40 ppm for gestating and lactating females compared to 20 ppm for all other classes of cattle (NASEM, 2016).

The Ca requirements for beef cattle range from 0.18-0.58% diet dry-matter (NRC, 1996), which would suggest that crabgrass alone would not meet the calcium requirements of all classes of beef cattle. The P requirements for beef cattle range from 0.16-0.26% diet dry-matter (NRC, 1996), meaning that P concentrations in this study exceed the P requirements of beef cattle. However, the present study did not evaluate P availability to the grazing animal. Phosphorus concentrations and availability can be impacted by several factors including environmental conditions, and stage of maturity (Dida, 2016). Potassium requirements range from 0.60-0.70% diet dry-matter with a maximum tolerable level of 3.0% (NRC, 1996). Potassium concentration in the present study ranged from 2.87-3.73%. However, the dietary requirements for beef cattle are based on total dietary intake. Although K concentrations in the present study exceed 3.0%, this does not mean that total dietary intake of grazing cattle would exceed the maximum tolerable level of 3.0%. The magnesium requirements for beef cattle range from 1.0-2.0% diet dry-matter with a maximum tolerable level of 0.4% (NRC, 1996). In the present study Mg concentrations 0.36-5.53%, however this does not mean that total dietary intake of grazing animals would exceed the maximum tolerable level. Another consideration for grazing beef cattle, is the interaction of high K concentrations resulting in Mg deficiency due to an inability of the animal to absorb Mg, this can result in the condition known as grass tetany (Suttle, 2010). In the present study, the high levels of K may partially be offset by high levels of Mg.

The Zn requirement for beef cattle is 30 ppm (NASEM, 2016), in the present study regardless of N fertilization rate or source crabgrass was at least marginally deficient in Zn, suggesting that supplementation is likely required. The manganese requirement for lactating and gestating females is 40 ppm, and is 20 ppm for all other classes of beef cattle (NASEM, 2016). In the present study, Mn concentrations exceeded the beef cattle requirement regardless of N application rate or source. Manganese concentrations fluctuate throughout the growing season (Suttle, 2010), but meet or exceed the requirements of grazing livestock. This should be considered when formulating mineral supplements, as to not over supplement unnecessarily. The Cu requirement for beef cattle is 10.0 ppm, (NASEM, 2016). In the present study crabgrass met the Cu requirement of beef cattle. However, concentration does not take into account availability of Cu in crabgrass. Thus, supplementation would still be recommended to insure adequate Cu intake. The iron requirement for beef cattle is 50 ppm (NASEM, 2016). Unlike other species of livestock, Fe supplementation is typically not recommended for beef cattle due to the high levels of Fe in most feedstuffs (NASEM, 2016). In the current study, Fe concentrations exceeded the requirements for beef cattle.

When evaluating forage mineral concentrations, it is important to understand that actual availability will likely differ from analysed concentration. Thus, simply because the concentration of a particular mineral within a forage exceeds the requirement of the grazing animal does not mean that the mineral may not also require supplementation to ensure that the requirements of the grazing animal are met.

The current study establishes mineral concentrations for crabgrass. For most minerals, crabgrass meets or exceeds the mineral requirements of grazing beef cattle. This coupled with previous literature evaluating the nutritive value of crabgrass, suggest that crabgrass could be a viable alternative for summer grazing of beef cattle. Future work should focus on understanding the availability of minerals in crabgrass and other forages to further refine recommendations for supplemental minerals.

Literature Cited:

Dalrymple, R. L. 1999. Crabgrass: A synopsis. p. 1-5. *In* Crabgrass for forage: Management from the 1990's. NF-FO-99-18. The Noble Foundation, Ardmore, OK.

Dida, M. F. 2016. Effects of Argo-Ecologies on Forage Species Quality. Res. J. An. Sci., 50(1-6): 1-5.

Gomez, K. A., Comes, A. A. 1984. Statistical procedures for agricultural research. 2nd. Ed. Hohn Wiley and Sons, New York, NY.

National Academies of Sciences, Engineering, and Medicine. 2016. Nutrient requirements of beef cattle. National academic press. Washington, DC.

National Research Council. 1996. Nutrient requirements of beef cattle. National academic press. Washington, DC.

Suttle, N. F. 2010. The mineral nutrition of livestock- 4th ed. Wallingford Oxfordshire. CABI Publishing.

Teutsch, C. D., Fike, J. H., and W. M. Tilson. 2005. Yield, digestibility, and nutritive value of crabgrass as impacted by nitrogen fertilization rate and source. *Agron. J.,* 97(6):1640-1646.

Virginia Department of Conservation and Recreation. 1995. Virginia nutrient management standards and criteria. Div. of Soil and Water Conserv., Richmond, VA.

Table 1. Mineral concentrations in crabgrass as affected by nitrogen rate and source. Fisher's protected least significant difference is shown in column 9 and 12 for N rate and N source, respectively.

	Nitrogen Rate (lb/acre)[a]								Nitrogen Source		
Mineral, %	0	50	100	150	200	250	300	NR LSD[b]	AN[c]	BL[d]	NS LSD[e]
Ca	0.28	0.27	0.29	0.33	0.32	0.37	0.39	0.049	0.36	0.29	0.026
P	0.41	0.42	0.41	0.37	0.35	0.34	0.32	0.026	0.35	0.39	0.014
K	0.29	0.31	0.32	0.36	0.36	0.36	0.37	0.040	0.33	0.34	0.022
Mg	0.37	0.36	0.41	0.44	0.45	0.50	0.53	0.040	0.44	0.43	0.022
Mineral, ppm											
Zn	27.0	24.7	26.4	25.0	25.1	24.6	25.1	2.04	24.7	26.1	1.09
Mn	94.4	78.7	85.1	105.4	105.6	107.0	136.7	18.8	135.9	67.8	10.0
Cu	10.6	10.9	10.8	10.3	10.2	10.1	10.1	0.896	9.91	10.8	0.479
Fe	62.1	56.0	58.1	59.7	65.8	63.2	63.6	5.71	62.4	60.0	3.05
B	4.45	5.08	4.73	4.73	4.50	5.20	4.53	1.71	4.21	5.28	0.916

[2a]Rate of nitrogen application (lb N per acre).
[b]NR LSD: Fisher's protected least significant difference for comparing N rates within a mineral.
[c]AN: Ammonium nitrate.
[d]BL: Broiler litter.
[e]NS LSD: Fisher's protected least significant difference for comparing N sources within a mineral.

AUTHOR INDEX